TALKING TECH

TALKING TECH

A Conversational Guide to Science and Technology

by

HOWARD RHEINGOLD
AND HOWARD LEVINE

WILLIAM MORROW AND COMPANY, INC.
New York *1982*

The authors would like to express their gratitude for permission to reprint from the following:

Haag, Jole. Poem from A *Random Walk in Science: An Anthology*, compiled by R. L. Weber, p. 138, 1974. Used with the permission of Crane, Russak & Company, Inc., New York, N.Y. 10017.

Wheeler, J. A. From *The Berkeley Physics Course, Vol. I*, edited by C. Kittel, W. Knight, M. Ruderman. Copyright © 1962. Used with the permission of McGraw-Hill Book Company.

H. M. K. "Fission and Superstition." January 14, 1950. Reprinted by permission of *New Statesman*.

Library of Congress Cataloging in Publication Data

Rheingold, Howard.
Talking tech.

Includes index.
1. Technology—Terminology. 2. Science—
Terminology. I. Levine, Howard, 1947–
II. Title.
T11.R44 601'.4 81-14130
ISBN 0-688-00783-X AACR2

Printed in the United States of America

First Edition

1 2 3 4 5 6 7 8 9 10

BOOK DESIGN BY MICHAEL MAUCERI

Acknowledgments

This book would still be no more than the focus of private conversation if it weren't for the help of people who believe, as we do, that the public's desire to talk tech reflects a fundamental shift in our perception of science and the purposes which it is to serve in our society. We gratefully acknowledge their help:

Dan Carpenter, our technical illustrator, and Larry Gonick, our cartoonist, who demonstrated that complex scientific ideas could be conveyed through simple visual means.

In order to insure that all the material in this book is scientifically accurate, we recruited a review panel of technical experts, representing a diversity of disciplines. While thanking them for their valuable suggestions, we accept full responsibility for any factual errors which may still remain: Dr. Mitch Baker, Department of Mathematics, State University of New York at Buffalo; Dr. Harvey Borovetz, Department of Civil Engineering and Surgery, University of Pittsburgh; Dr. Alphonse Buccino, National Science Foundation; Dr. Chris Cherniak, Department of Philosophy, Tufts University; Dr. David Krohne, Department of Biology, Wabash College; Dr. David Lichtenstein, Department of Psychology, State University of New York at Stony Brook.

During the nearly three years it has taken to turn these ideas into a book, we have received help, ideas, and encouragement from Rita Aero, Gerri Deckter, Judith Maas, and Dr. Alex Morin. We wish to thank them as well.

Those who draw a distinction between education and entertainment don't know the first thing about either.

—MARSHALL MCLUHAN

We have a society which is built on science and technology and which uses science in every one of the interstices of national life, and in which the public, the executive, the legislative, and the judiciary have very little understanding of what science is about. That is a clear disaster signal. It has to be suicidal.

—CARL SAGAN

PREFACE
Talking Tech—A New Social Requirement

The age of innocent faith in science and technology may be over. . . . every major advance in the technological competence of man has enforced revolutionary changes in the economic and political structure of society.

—BARRY COMMONER

People must understand that science is inherently neither a potential for good or for evil. It is a potential to be harnessed by man to do his bidding. Man will determine its direction and its effects. Man, therefore, must understand science if he is to harness it, to live with it, to grow with it.

—GLENN SEABORG

Talking tech—the ability to converse knowledgeably about scientific and technological topics—has become a new social requirement. This has occurred, in part, because everyone seems to be doing it. People can be found talking tech just about everywhere: abortion is argued at bus stops, clone jokes are traded at beauty parlors, food additives are the talk of the PTA, and genetic engineering is hot gossip when stockbrokers gather. Everyone seems to spice his small talk with words like microprocessor, black hole, and CAT scanner. Even scientists and engineers are having trouble keeping up with the technological topic of the week.

As a popular social pastime, the ability to talk tech puts a premium on entertainment value. Nobody wants to sound like a textbook when he makes casual conversation, so it isn't sufficient to understand the principles behind a new technology—you also have to keep your listeners interested. Because of this dual requirement—to *convey* as well as comprehend—*Talking Tech* is designed to be as much a fun house as a classroom. Here is the information textbooks normally leave out—cartoons, demonstrations, quizzes, poems, anecdotes, epigrams, jokes, and amusing tales to tell the next time somebody mentions cosmology at a party or brings up meltdowns at the bridge table. When Johnny asks where the energy of the future will come from, or your date expresses an interest in reproductive technologies, *Talking Tech* will not only enable you to give a lucid explanation, it will also provide you with conversational ammunition.

Tech talk is far more than a passing fad or intellectual recreation. People are conversing about science not only because it's the current fashion but because a genuine social transformation is taking place—a fundamental shift in the relationship between the larger society and the scientific community. This shift in mass attitude isn't anything that was planned in advance; it just seemed to happen to us. The news media were partially responsible, but newspapers and television only transmitted the news of the events. History itself gave us such surprise science lessons as Hiroshima, thalidomide, and the Skylab crash. Science has been pulling us all along into uncharted territory at a dizzily accelerating rate. And something has been happening to our minds along the way: Our attitudes toward science and technology have been changing. It started to happen somewhere between the Salk vaccine and the Love Canal, sometime after Atoms for Peace but before the monster from Three Mile Island. One morning we awoke to the thought that perhaps scientific progress wasn't our most important product after all.

Although we may not be consciously aware of it, the general population is always involved in an unwritten contract with the subpopulation of scientists. Like all contracts, this deal is not immutable. Occasionally the entire agreement is dissolved and renegotiated. It happened in ancient Greece, in Renaissance Rome, in nineteenth-century London, and now it is happening in twentieth-century America.

Just as science itself proceeds in a discontinuous process of crises and revolutions, the social relationship between science and citizenry also travels a bumpy road. If science didn't fare too well in the trial of Galileo and came in for some hard knocks from the pen of Dickens, its path through this century has generally been a smooth one. In fact, many claim that science and technology have been elevated to such a lofty social altitude that they have become quasi-religions. But just when we began to feel comfortable with this relationship, our new god started doing weird things to us.

After *Skylab* fell from orbit and everyone on earth played an involuntary round of cosmic Russian roulette, we all began to wonder what *else* might happen. Three Mile Island created even more of a furor. But those of us who weren't in the Australian outback when the satellite crashed, or who didn't live near Harrisburg, Pennsylvania, soon found more urgent problems to solve in our everyday lives. The almost weekly stories about leakages from unsuspected toxic waste dumps and the profusion of reports regarding cancer-causing agents in our food, water, and air made it seem that we were no longer safe even in our own homes. When you discover that your backyard is situated over a chemical dump, or that a substance you've been working around for ten years is a suspected carcinogen, it's hard not to take a personal stand on scientific issues.

Science concerns us more and more these days because it is becoming more and more involved in our daily lives. It is intruding on our ideas of who we are. Thanks to our successful probes into the nature of the universe, the human place in the scheme of things is being challenged once again. What is it going to mean to us as individuals and members of the human race when biologists create life, when cyberneticians build a mind, when exobiologists contact extraterrestrial intelligence, or when chemists synthesize a true anti-aging drug? Judging from the scientific shockers of the last twenty years, we should all be apprehensive about what might happen to us over the next two decades.

This anxiety, or at least a vague uneasiness, is the root cause of a new social requirement: the ability to talk tech. Until very recently, we gave science and technology a sweet contract, virtually a blank check. The researchers and engineers promised to make our lives whiter, brighter, and fingertip convenient, and we promised not to ask too many questions about those chemicals in our breakfast cereal. The new contract requires much more mutual involvement and understanding. The current dispute in America over the value of science in our society stems not simply from a loss in confidence in scientists or from a decline in the belief that science can help solve most of our problems. It is fundamentally a dispute about how much influence sci-tech should have on our lives and who should be making decisions about the extent of that influence.

Much has been written lately about the public's antiscientific attitude. This appears to be a misinterpretation of the evidence. The public has not turned against science, but toward responsible science. In a sense, science has become too important to be left solely to the scientists. We are beginning to perceive that the best way to insure greater responsibility is through our own participation, and conversation is an important first step toward participation. *Talking Tech* furnishes the backup knowledge you'll need to take that all-important step with confidence and aplomb.

In the next days and decades, we will all have to make personal decisions about energy alternatives, the transportation/storage/disposal of dangerous substances, microwave hazards, test-tube babies, synthetic life forms, euthanasia, life-extending technologies, the confidentiality of computerized personal data, and a dozen other science-based subjects we haven't even heard about yet. For better or worse, these scientific developments are going to change our lives, and the ability of nonscientists to participate in making those changes depends on the quality of information communicated to us and among us.

The public demand for scientific information is reflected in the recent explosion of popular science media—magazines, television series, books, newspaper supplements. The science bubble in the mass media is not a fad or a scam, but the manifestation of a widespread need for understandable explanations of the latest sci-tech events. People want to know about science. Perhaps more importantly, they want to show other people that they know. It may be true that a social contract is being rewritten, but most of us just want to explain how our new microwave oven works or be able to debate nuclear power at the neighborhood tavern.

Talking Tech is designed to be a decoding device to help you decipher the esoteric jargon of science and technology. In "The Fundamentals," you will be introduced to the general strategy of tech talk: how to talk precisely without becoming encumbered with mathematics, how to figure out which topics belong to which science, and even how to use some of the euphemisms scientists use to fudge their own uncertainties. In "The Specifics," everything from acid rain to Zeno's paradox gets the full treatment: a short precise definition is followed by a longer, easily readable explanation of what it *really* means, supplemented by a section providing conversational tactics, a brief cross-listing of related topics, and a short list of further readings. The epilogue is a tech-talk crossword puzzle, an opportunity to prove that education and entertainment go hand in hand.

Tech talk will be an increasingly significant social dialect in the years to come. *Talking Tech* is meant to help you meet the challenge of speaking it head-on.

Contents

INTRODUCTION
Talking Tech— The Fundamentals

Science is a first-rate piece of furniture for a man's upper chamber, if he has common sense on the ground floor.

—OLIVER WENDELL HOLMES, SR.

In good speaking, should not the mind of the speaker know the truth of the matter about which he is to speak?

—PLATO

Talking tech is what the public relations officer from the chemical plant is doing when he speaks of polychlorinated biphenyls in the food chain, and what you really want to know is whether that stuff oozing from the ground will turn your kids into mutants. When your doctors gather around your bed and exchange polysyllabic mystery words, they are also talking tech. Talking tech is what the newscaster is doing when she announces that your neighborhood is thirty millirads hotter since the latest venting of the local nuclear reactor. Even though technology is often threatening, tech talk isn't necessarily a malignant jargon: it is what your brother engages in with his friends down at the computer store, and it's what the PTA is doing when abortion, evolution, or food additives are debated. In fact, so much tech talk goes on around us all the time that we are constantly receiving lessons in the special language of science. This introduction is designed to help you make the most of these free lessons, to make you a better listener as well as a better tech talker.

We swim in a sea of specialized dialects. Americans seem to relish talking football ("They play a flex with a stacked, rotating zone") or talking stereos ("I prefer electrostatics with a vacuum-tube amp and moving coil cartridge") or talking taxes, or cars, or CBs, or skiing . . . . Like any of these languages-within-languages, talking tech requires a special vocabulary as well as a knowledge of the grammar needed to manipulate those strange words. However, unlike these other tongues, the ability to talk about science and technology is akin to the ability to actually do science, because communicating about science is the essence of science itself. Talking tech starts with the art of asking questions; science is basically a system for asking questions and then publicly debating and explaining the answers.

Science is not so much an ever-changing body of knowledge as it is a process for accumulating knowledge—a self-transforming, dynamic interplay between observation and theory. *Observations* are a way to compare our experience of the world, to open windows between individual minds and their shared reality. Ideally, observations are constant, unchanging perceptions that can be experienced by any observer in the same circumstances. Anyone who looks at the sky or the sea can use his senses to confirm the motions of the stars, the progression of the tides.

Theories are ways of explaining observations, of drawing as many public perceptions as possible into the same explanatory structure. If observations are the vocabulary of science, theories constitute its grammar. There are strict tests for observations to correct for individual bias; likewise, there are strict tests for theories. A good theory is one which provokes new questions and new kinds of observations. Once an explanatory mechanism is found for observed phenomena, the theory is not truly complete until it dares to forecast the future, to predict a specific kind of observation which had never before been reported. In the interplay between theory and experiment, question and observation, prediction and explanation, is the element which gives life to the whole process—the *conversation* of science.

Science, most particularly, is a *public* body of knowledge, a realm of discourse open to anybody who has something valuable to say. When people make scientific discoveries, whether it is in a basement laboratory or a giant research facility, they publish their findings for everybody else to argue about. It doesn't matter who you are—what matters is that your ideas fit together facts with new predictive or explanatory power. When a junior patent clerk in Zurich claimed to be the first person to truly understand the nature of space and time, all the eminent physicists in the world argued about his bold assertions, checked his equations, performed a few crucial experiments, and eventually agreed that young Albert Einstein did indeed know what he was talking about.

Knowing what you are talking about is more than a rule of the conversational game. Discourse, discussion, dialogue, and debate are the living breath of the scientific process. One person can be a scientist, can perhaps create an entire discipline, but it takes many people, communicating their ideas over many years, to build up a self-consistent body of knowledge. As Newton said regarding his own remarkable insights, every scientist "stands on the shoulders of giants."

The boundaries of the conversational domain we have been calling tech talk are by no means exact, nor could they be. Like science itself, the subject matter of tech talk is always shifting, adding new topics or discarding old ones with each scientific discovery (e.g., quasars) and every perceived social implication (e.g., computers and personal privacy). Exobiology may be elevated from a strictly theoretical to a shockingly observational science, or fiber optics may cease to be a topic of basic research and become a widely used technology. But although the specific entries themselves might change, the basic conversational structure remains the same.

The language of science is a framework for comparing facts and ideas, observations and theories. By learning the language, you learn how to construct your own conversational frameworks. The way to learn any language is to learn how to listen to those who speak it. Obviously, the first step is to recognize when somebody is talking about science. The next step is to determine what *kind* of science they are talking about. New topics may surface every day, but most of them are related to existing categories of discussion.

The conversational constants are the categories of scientific conversation, and they provide the first rule of talking tech: *Science is not a monolithic discipline, so be prepared to talk about the subspecialties within the sci-tech domain.*

The phrase "science and technology" encompasses so many subcategories that it must be understood as a very broad classification, a kind of galactic cluster of related concepts, many of which have planetary subsystems of their own. When you hear somebody discussing science or technology, he could be talking about a scientific *pretender* (common sense, nonscience, prescience, or pseudoscience), a scientific *cognate* (research, development, engineering, or technology), or a scientific *qualifier* (applied, basic, experimental, or theoretical). In order to broadly map the territory covered by the seventy specific topics, *Talking Tech* provides this capsule guide to the categories of scientific discussion.

Scientific Pretenders

Common Sense: Science begins with common sense (roughly defined as what we have learned on the basis of everyday experience), but it pushes into domains where common-sense investigation cannot go, such as the realms of the very small, very large, and very fast. While exploring these zones beyond the borders of everyday experience, science may even end up proving common sense to be *wrong,* as in the seemingly paradoxical results of quantum theory and relativity theory. In a peculiar way, the science of yesterday becomes the common sense of today.

Nonscience: Essentially, the distinction between nonscience and science mirrors the difference between politics (how to get elected) and political science (how to govern). Although no simple set of characteristics distinguishes scientific from nonscientific disciplines, a few traits are most closely associated with science: the explanation, prediction, and control of natural or social phenomena are the hallmarks of true science. The development of new tools, techniques, or theories may change a nonscience into a science, as seems to be happening with economics (see the chapter Mathematical Modeling).

Prescience: This is an approach to a problem which may produce sound results but is not as systematic and complete as science. Scientists distinguish between the context of discovery and the context of justification. In the former, anything goes. For example, inspiration for the bubble chamber came from a glass of beer (see Particle Accelerator). In the latter, results must be proven according to the rigorous rules of scientific method. Just because certain tribes successfully treated illness with herbs or ancient civilizations were able to capitalize on mutant wheat strains doesn't mean they practiced *science.*

Pseudoscience: This embodies an approach to a supposedly scientific problem which is heavy on the jargon but light on testable explanations. Topics such as the Bermuda Triangle, ancient astronauts, and pyramid power all fall into this category. If in doubt, ask yourself two questions: Is there a *simpler* explanation of the facts? And what thoroughly proven scientific data do I have to give up to accept this new belief? Although pseudoscience sometimes becomes real science, as appears to be the case with acupuncture, more often than not it remains a kind of intellectual sleight of hand.

SCIENTIFIC COGNATES

Research: The Holy Grail, what they award the Nobel Prize for, and what most of us imagine when we think about science, research is the systematic inquiry into why the world works the way it does. Although research used to be considered an unadulterated good, performed by white-coated scientists to benefit us all, recent scandals concerning the protection of human subjects (see Bioethics), as well as concerns about the safety of some kinds of research (see Recombinant DNA), have brought research under as much scrutiny as its running mate, technology.

Development: The "D" of "R & D." It is the link between a scientific idea (energy from atoms) and a prototype of that idea (the fusion reactor). Because hardware is more expensive than ideas, two thirds of the more than fifty billion dollars the U.S. spends yearly on science goes into development (see Fusion).

Engineering: This is usually thought of as science's poorer cousin. (Quick, name two famous engineers. You might name Herbert Hoover and Jimmy Carter, but they are hardly famous for their engineering achievements.) Engineering is the practical application of scientific principles. It is the art of taking a successful development project like ENIAC (Electronic Numerical Integrator and Calculator) and using that new knowledge to build a whole "generation" of digital computers (see Miniaturization). In a sense, engineering disseminates science's findings to the rest of us in concrete form.

Technology: If it weren't for technology—the production or transformation of material objects and the creation of procedural systems—there would be no need for *Talking Tech.* Technology not only changes our lives, it changes the way we think and talk about our lives. It has this effect because it extends our human faculties (see Image Enhancement), increases our efficiency (see Fiber Optics), and reduces risks (see Technology Assessment). Unfortunately, it also has the capacity to affect our lives negatively (see Acid Rain, Electronic Smog, Toxic

Chemicals), and it is precisely this intrusive aspect of technology which has spurred the new public interest in sci-tech.

SCIENTIFIC QUALIFIERS

Applied Science: The key notion here is that science can be immediately focused to try to solve a real-world problem, rather than proceeding by its usual scatter-gun approach. Sputnik signaled the beginning of our major efforts in applied research and led to a whole "scientific-solutions-to-order" mentality (Energy Alternatives). The space program's early successes also led to heightened expectations best expressed by the line, "If we can put a man on the moon, then why can't we [end poverty, rebuild the cities, cure cancer . . .]."

Basic science: Pure, or basic, research is thought to be the driving engine behind all scientific progress. As opposed to the solution-oriented restrictions of applied research, it is a scientific exploration into the ways of the world for knowledge's own sake, or, as Wernher von Braun said, "Basic research is what I am doing when I don't know what I am doing." Classic examples of basic research include solving the scandal of the ultraviolet (Blackbody Radiation) and determining the molecular structure of the genetic code (DNA).

Experimental Science: Mark Twain said, "There is something fascinating about science. One gets such wholesale return of conjecture out of such a trifling investment of fact." It is the job of the experimentalist to see that the number of facts (the data of science) is constantly growing. Experiments are the method by which scientists confirm theories as well as discover phenomena which require theoretical explanation. When scientists split atoms (Particle Accelerators), or merge the genes of different organisms (Recombinant DNA), they are engaging in experimental research.

Theoretical Science: "Theoretical" is science's glamour qualifier. Theories—comprehensive sets of statements which can explain and predict natural phenomena—are considered to be science's highest achievement (Periodic table). One way theories explain phenomena is by employing concepts such as consciousness, which are known as theoretical terms because they have no single obvious observational referent. In the world of science, the creed of the experimentalist is "Seeing is believing," while the motto of the theoreticians is "I'll see it when I believe (or explain) it."

Getting a Grip on the Numbers

Once you've mastered the ins and outs of S & T and R & D, it's time to move on to an essential aspect of technical discussion—quantification. As science pushes into domains which are further and further removed from our everyday experience (distances measured in trillions of miles and time counted in billionths of a second), it becomes more difficult to talk tech accurately. Everett Dirksen, the onetime House minority leader, used to say: "A billion here, a billion there, and pretty soon it adds up to real money."

The problem is much the same in talking about modern science and technology—the numbers are so big, or so small, that it is easy to lose the feel of reality, to miss the mark by a factor of 100 (Does light cross an atom in 10^{-18} or 10^{-16} seconds?) without even realizing it. The easiest way to avoid this number-vertigo is to focus on what scientists call *degrees of difference* or *orders of magnitude.* Differences of the powers of 10—how many times you can multiply or divide the number by 10—are the way most scientific quantities are discussed.

If your child is ten and your mother is one hundred, there is one degree of difference, or order of magnitude, between them. Nobody is born with an intuitive understanding for the meaning of 10^{27} (1 with twenty-seven zeros after it) or 10^{-13} (a decimal point followed by twelve zeros and a 1), but you can get a feel for the numbers by comparing them to their neighbors. A sense of scale is more important than a memory for figures when it comes to talking about science.

The second rule of talking tech: *Try to make quantities comprehensible by using examples and comparisons; look for ways to relate incomprehensible quantities to human scale.* The following illustrations can help you locate your topic of discussion in the proper space-time framework.

You Can't Tell the Players Without a Scorecard

Since the end of World War II, the ability to talk tech has required more than the ability to talk about relativity or acid rain or any other purely scientific or technological topic. We are now living in what the experts call the era of Big Science—big dollars, big equipment, big laboratories, and big bureaucracies. When Drs. James Cronin and Val Fitch won the Nobel prize for physics in 1980, they shared the honor with their technicians, their graduate students, their research assistants,

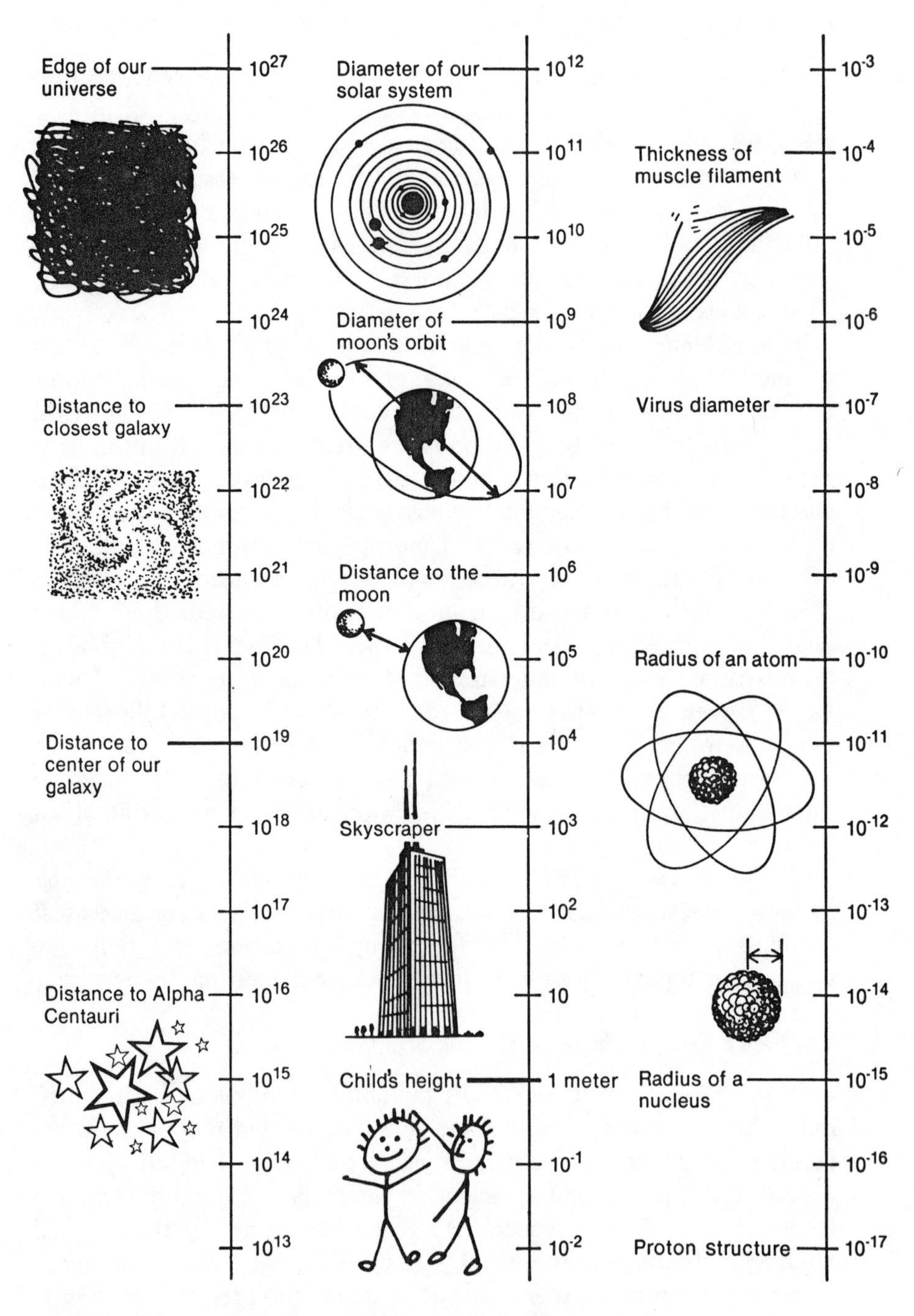

The Fundamentals—Cosmic Space Zoom

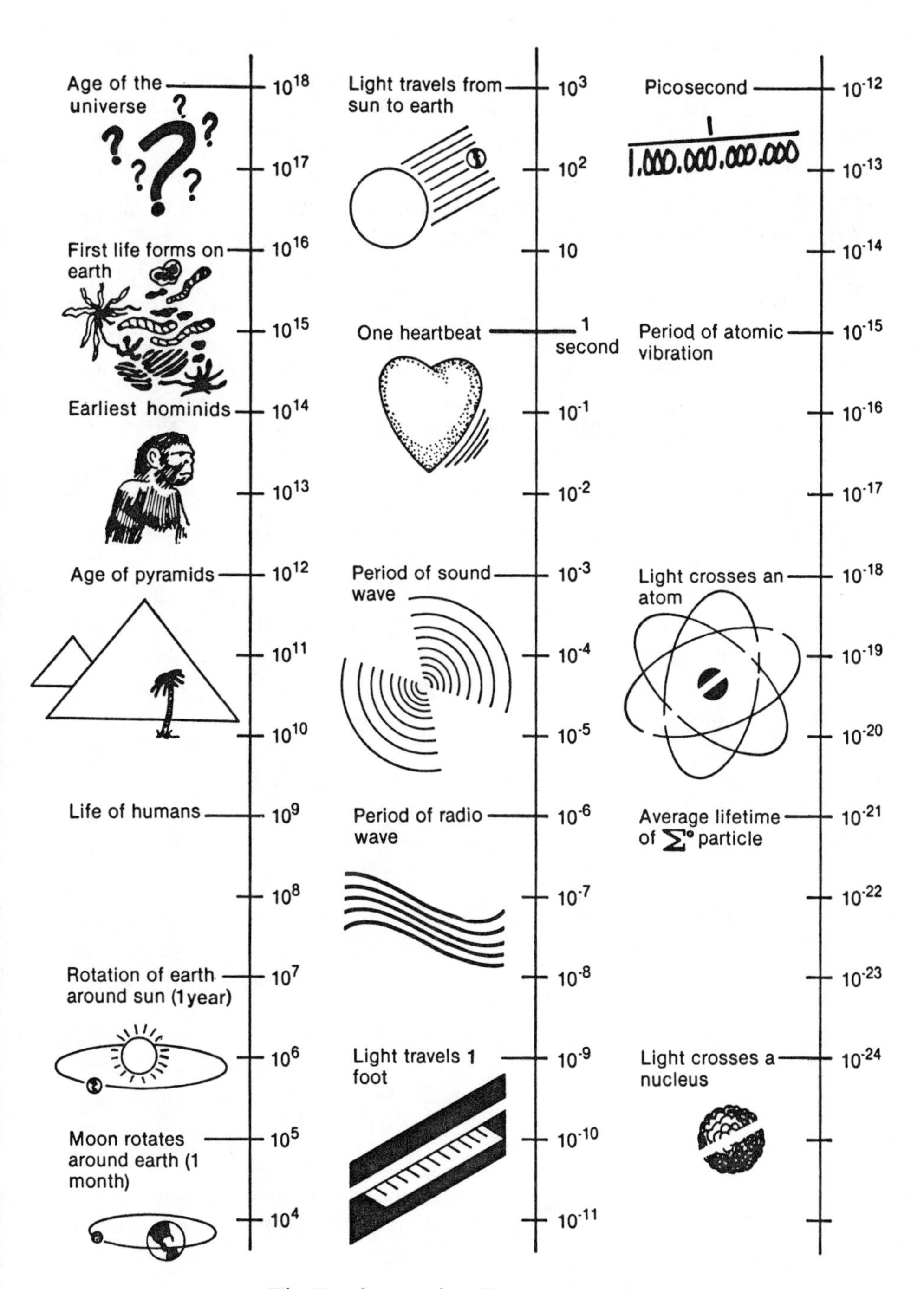

The Fundamentals—Cosmic Time Zoom

their postdoctoral fellows, and perhaps most importantly, their institutional grants officers and their funding agency's monitoring officer. Science is no longer conducted by lone figures dressed in tattered white coats and working in ill-equipped little laboratories.

Scientific research and development is big business: the total U.S. bill for R & D is about twice the total budget for the state of California, and that requires big institutions to control it. In fact, most of us never get to see, hear, or talk to a working scientist. Instead, we receive our explanations from the chemical plant's PR man, or a spokeswoman for the laboratory, or a government press officer, or some type of administrator-cum-scientist. This buffer between scientists and the lay public will continue to grow as science becomes more political and more involved with the making of public policy choices.

This politicization and bureaucratization of science will also force us to become more aware of the giant science and technology organizations already existing in our society—and the roles they play in our lives. In order to get a jump on your conversational cronies and help get the government working for you, *Talking Tech* presents this guide to science's "alphabet soup." The third rule of talking tech is: *The successful practice of science today requires political as well as technical skill, so be prepared to talk about the political aspects by knowing which agency is responsible for what tasks.*

SCIENCE'S ALPHABET SOUP

Organization	*What They Do*
AAAS—American Association for the Advancement of Science	The nation's largest public-interest science group; they publish the magazines *Science* and *Science 82.*
DOD—Department of Defense	With the government's largest R & D budget (about twelve billion in 1980), they are into everything from MX missiles and killer satellites to computer-assisted instruction and better boots.

DOE—Department of Energy	The name tells it all—they handle everything from nuclear power to more efficient light bulbs.
EPA—Environmental Protection Agency	A key agency when it comes to public policy matters. If you have a complaint about toxic wastes, acid rain, or carcinogenic substances, tell it to them.
FDA—Food and Drug Administration	The folks responsible for seeing that we don't ingest unhealthful substances. They are constantly making headlines by banning another drug (Laetrile) or food additive (saccharin).
NAS—National Academy of Science	A peculiar quasi-governmental agency established by President Lincoln to provide the government with scientific advice. Admission is only by election (a kind of Hall of Fame for scientists).
NASA—National Aeronautics and Space Administration	Perhaps our best-known science agency.
NIH—National Institutes of Health	A conglomeration of research laboratories (National Cancer Institute and National Heart, Lung and Blood Institute are the two largest) which spends about three billion dollars a year researching disease.
NOAA—National Oceanic and Atmospheric Administration	A division of the Department of Commerce which serves as the national weatherman and marine biologist.

NSF—National Science Foundation.	A billion-dollar agency which is chiefly responsible for our nation's conduct of basic research.
OSTP—Office of Science and Technology Policy	An office within the White House charged with giving the President advice regarding science and technology matters.
OTA—Office of Technology Assessment	A congressional office charged with giving Congress advice on science and technology matters.
RAND—Formerly Research and Development Corporation	A think-tank which started as the Air Force research and development office and is now a private, non-profit corporation, conducting studies for the government and private clients.
SRI—Formerly Stanford Research Institute	The country's largest private think-tank.

Meaning What You Say and Saying What You Mean

Science is known for its precision. One would assume that there is one major difference between talking tech and talking anything else—the accuracy with which sci-tech matters must be discussed. As it turns out in practice, this is true only part of the time. Any great theorist could tell you that it often takes a thousand farfetched speculations to create one plausible hypothesis. Scientists, not wanting to let on that they fudge as much as the rest of us when it comes to tech talk, have developed their own little euphemistic code.

As your next step in becoming an accomplished tech talker, you should learn how to sound scientific without knowing exactly what's going on. Familiarize yourself with table following and remember rule four of talking tech: "*What I say isn't always what I mean*" *applies as much to science as it does to politics, so be aware that it's all right to admit ignorance if you use the proper code words and quickly move to the interesting part of the story.*

What You Say	*What You Mean*
Everyone knows that . . .	I don't have a precise reference.
Preliminary experiments indicated . . .	I heard it mentioned once.
Of great theoretical and practical importance.	Interesting to me.
Correct within an order of magnitude.	Probably wrong.
Some researchers believe . . .	Nobody knows for sure.
Typically, the results are . . .	The best results are . . .
It might be argued that . . .	I can't refute this objection, so I'll raise it myself.
The implications are obvious.	My bias is clear.
It is generally believed that . . .	A few other people besides myself think that . . .
Will soon be in every home.	I saw some ad that claimed 100 percent market penetration by the year 2000.

With a solid grounding in the fundamentals of tech talk, you should now be prepared to move on to the personally interesting, conversationally potent topics that make up "The Specifics" of *Talking Tech.*

Talking Tech–The Specifics

The two most engaging powers of an author are to make new things familiar and familiar things new.
—WILLIAM MAKEPEACE THACKERAY

Everything should be made as simple as possible, but not simpler.
—ALBERT EINSTEIN

The most important topics to have in one's tech-talk repertoire will always be a matter of some debate. This is the result of the individual's sense of what really matters, as well as the constant accumulation of new sci-tech information. After a careful review of the scientific and popular media of the past decade, we decided that the following seventy topics were the most important to the most people. If we missed your favorite subject, check the index. The odds are good that we've discussed it within another topic.

The *Talking Tech* specifics fall roughly into three classes. First, the hardy perennials: These are topics, such as the periodic table or the theory of relativity, which have been essential knowledge for decades and will continue to be essential because they form the context, the explanatory framework, for the whole field of talking tech. Second, the recent bumper crop of "most-talked-about" subjects: These are topics, such as acid rain, energy alternatives, fiber optics, and recombinant DNA, which were virtually unknown ten years ago but which are now nearly household words. The third category concerns the seedlings of future discourse: These are subjects, such as electronic smog, longevity-immortality, or intelligence enhancement, which are currently being discussed only among sci-tech experts but which hold the promise of soon becoming popular topics of conversation.

Each of the seventy topics is given the same, five-part treatment:

DEFINITION:

A brief, succinct definition of the type usually found in scientific dictionaries. Memorizing the definition will enable you to do crossword puzzles or nod in assent if the topic is mentioned, but it won't do much to help you talk about the subject.

WHAT IT REALLY MEANS:

A concise explanation of the topic, including why it is important, how it works, its historical antecedents, and its impact on daily life today—in plain English. Images, metaphors, and models are used when appropriate.

CONVERSATIONAL TACTICS:

Since talking tech implies *talking,* not just understanding, this section contains tips to help you communicate what you know. It offers anecdotes, basic information, controversies, speculations, jokes, poems, demonstrations, quizzes, experiments, and useful comparisons to enliven your tech-talk style.

RELATED TOPICS:

In science, it's axiomatic that everything is connected to everything else; however, some topics are more connected than others. In this section we have tried to select, for each topic, the half dozen or so most closely related subjects, to enable you to immediately cross-reference other topics if you desire more information.

FURTHER READINGS:

These shouldn't be interpreted strictly as references for the preceding material, but as aids to your own further investigation. In the course of researching *Talking Tech*, we reviewed scientific reports, unpublished government documents, and articles in technical journals, as well as more popular material. Inclusion under this category was determined by two criteria—quality and accessibility.

ACID RAIN

DEFINITION:

Any form of precipitation having a pH less than 5.5, the reading for normal rainwater. It is produced by a combination of natural meteorological phenomena and human-created atmospheric pollutants.

WHAT IT REALLY MEANS:

"As pure and gentle as rainwater" was, at one time, an effective advertising slogan—but not any longer. In fact, if you live in the Northeast U.S. and put a cruet of olive oil outside in a particularly noxious rain, you might just end up with a rather unpalatable vinegar-and-oil salad dressing. Unfortunately, this unsavory result is neither the sole nor the most serious consequence of acid rain: the statues in the Parthenon have almost crumbled, many lakes in New York's Adirondacks have "died," and in all areas of the world vegetation is suffering a doubly deleterious effect—the foliage is assaulted from above, while the roots are poisoned and starved in the soil below. It's not just your Aunt Minnie's prize roses that won't bloom; crops all around the world are seriously threatened.

Like many ecological threats, acid rain is easy to define—it is simply any form of precipitation more acidic than normal rainwater. Discovering how to curtail it is another matter. Nor is the problem strictly technological. One major social and political problem is that the kind of atmospheric pollutants which cause acid rain are exactly those by-products of our petroleum-dependent society that are bound to *increase* in the energy-crunched years ahead. The diagram on the opposite page contains a simplified explanation of the mechanism of acid rain.

Halting the effects of acid rain is also proving to be more difficult in practice than it is in theory. Schemes such as liming acidified lakes, breeding acid-resistant fish, and using extra-tall smokestacks have all been tried and discarded. Using smokestack scrubbers to remove much of the sulphur at the point of emission has provided the best results thus far; however, they are very expensive and don't get rid of the nitrogen oxides and some of the other pollutants. The current easing of auto emission and coal-burning standards in response to the rising price of petroleum will also exacerbate the condition.

Because of these problems, it is not too surprising that acid rain has been called perhaps the most serious environmental dilemma in this century. When one looks at some of the other topics in this book, from meltdowns to toxic chemicals, it becomes clear that such a claim carries ominous weight. Acid rain is but one harbinger of the coming age of technological/social trade-offs.

CONVERSATIONAL TACTICS:

The garden is a perennial topic of conversation in the suburbs, and the condition of the crops serves an even wider conversational function in farm country. "What did you plant this year?" "How are you coping with mealy bugs?" "Did you see the size of the tomatoes Rheingold harvested this year?" Everyone should have one good piece of gardening advice for those occasions when the talk turns agricultural. Besides, acid rain is the newest twist on the oldest known small-talk topic—the weather. As a service to weekend gardeners, and others who just want to be able to converse about it, here is *Talking Tech*'s capsule guide to acid rain in your own backyard.

The key is to stay ahead of the problem. Once your soil starts to become acidic, it becomes increasingly more difficult to restore. Collect rain samples in your garden and use litmus paper or a soil-testing kit to determine the pH levels of the rain. Inspect your plants for spotty signs of damage in places where rain collects, because sulfates attack plant cells. You should determine if those of your plants which are "acid-

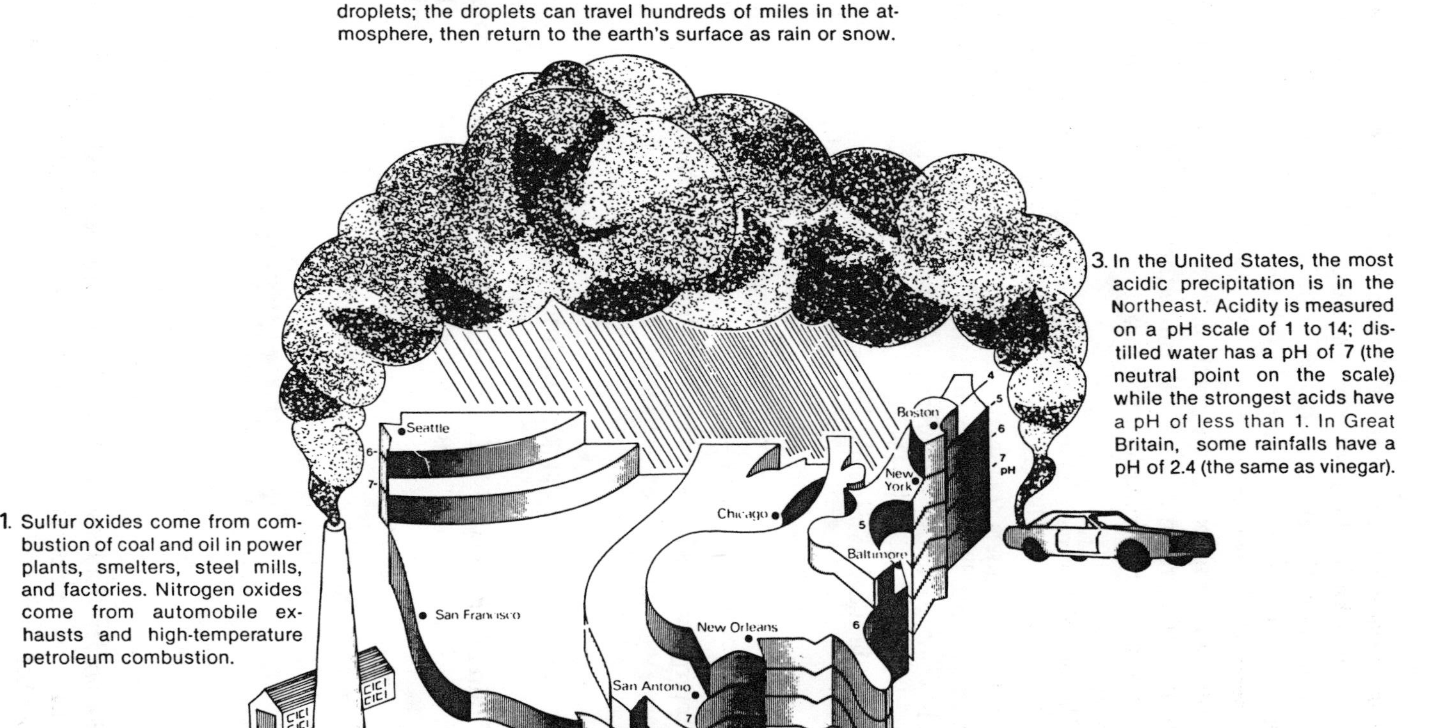

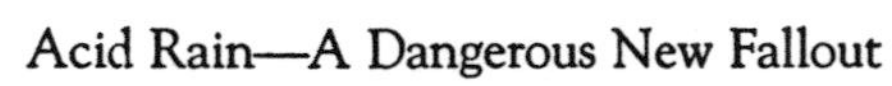
Acid Rain—A Dangerous New Fallout

loving" are doing a great deal better than those which need alkaline soil.

Once you know the nature of your backyard microclimate, you can take a number of steps if acid rain is indicated. Eventually, the rains affect the soil, so you must remedy this with a soil antacid: bone meal, powdered potash, lime, or wood ash. Nutrients such as potassium, calcium, and magnesium are leached from acidic soils, so you should consider adding them to your soil. Since the acidity also makes it tough on the earthworm population, you might let a few dozen of these helpful critters loose.

Ultimately, you may have to consider landscaping with "acid-loving" plants: azalea, blackberry and other brambles, blueberry, chrysanthemum, conifers, cranberry, marigold, oak, peanut, pecan, radish, raspberry, rhododendron, watermelon. If you have the opposite problem you'll want to know the alkaline-loving plants: alyssum, asparagus, bean, beet, cabbage, cantaloupe, carnation, cauliflower, celery, cucumber, iris, lettuce, nasturtium, onion, pea, phlox.

RELATED TOPICS:

Appropriate Technology, Energy Alternatives, Photosynthesis, Technology Assessment, Weather Modification.

FURTHER READING:

"Rain of Troubles," C. K. Graves, *Science 80*, July/August, 1980.
"Acid from Heaven," Susan West, *Science News*, February 2, 1980.

ACUPUNCTURE

DEFINITION:

An ancient Chinese method for relieving pain and inducing a variety of cures by inserting long, thin needles into specific body points just under the skin.

WHAT IT REALLY MEANS:

Prior to Ping-Pong diplomacy, anybody who asked his doctor to impale him with needles, even for medical purposes, would have been considered something of a kinky masochist. Then Nixon went to China, journalist James Reston had a well-publicized appendectomy in China using acupuncture anesthesia, and all of a sudden becoming a human pincushion became the thing to do. Acupuncture is no longer faddish and is still resisted by the AMA, but it is becoming better understood and more clinically accepted by some Western physicians.

Although the Chinese claim that acupuncture can treat nearly three hundred different diseases, from asthma to ulcers, Western interest is currently focused on the use of acupuncture as a painkiller. This is because acupuncture analgesia has attractive advantages over the drug-induced semicoma used in most hospitals today: There is no worry about drug levels or side effects; there is no sedation-heavy postoperative period; and the analgesic effect of acupuncture lasts for several hours after the operation. Why, then, isn't acupuncture an option on your Blue Cross card?

The major reason is that Western doctors, for good or ill, are notorious skeptics. Sometimes seeing isn't necessarily believing; in medical research, *explaining* is believing. Without a physiological explanation for the mechanism for acupuncture, most doctors would just as soon invoke voodoo. In fact, the time-honored Chinese explanation makes about as much sense as voodoo to Western scientists: There are two polar forces in our body, the yin and the yang, which combine to form a kind of life force called *chi.* This life force, according to ancient Chinese medical texts, flows through the body along a series of channels called meridians.

When the yin and yang fall into disharmony, disease and pain occur. When acupuncture needles are inserted at specific points on the meridians, the two forces are brought back into harmony. The fact that these channels do not correspond to known anatomical structures has long been a stumbling block to the medical establishment, and all the vague talk of life-forces doesn't sound scientific enough to someone who has spent eight years studying neurotransmitters and galvanic potentials.

Unfortunately, the standard Western model of pain and analgesia—the specificity theory—offers little more by way of explanation for how acupuncture anesthesia works in practice. According to this model, specific pain receptors in our bodies relay their signals directly to the brain, like a primitive telephone switchboard—stub your toe and the signal reaches the brain, often triggering an "ouch" response from your vocal cords. This theory has two logical consequences: Pain is felt precisely at the point of stimulation (if you bash your thumb with a hammer, your thumb and not your ear is where you feel the pain); the amount of pain experienced also depends on the intensity of stimulation (how hard you were swinging that hammer when it intercepted your thumb determines how loud your "ouch" is). Far from explaining it, this theory also contradicts the observed effects of acupuncture anesthesia. The specificity theory simply cannot account for how a needle in the earlobe or ankle can make the pain of angina disappear.

Fortunately for the advocates of acupuncture, other anomalies have turned up which the specificity theory also fails to explain: Referred pain, a well-known symptom in which cardiac patients often develop pain in the left arm or shoulder, is evidence that there are neurological links between distant body parts. Work with amputees who suffer from phantom-limb pain indicates that brief, mildly painful irritation at the perceived source of pain (the phantom limb) often brings about substantial relief. The specificity theory cannot explain how additional pain can lead to relief. Finally, there is experimental evidence that one pain can produce a marked decrease in sensitivity to other types of pain. All of these problems, as well as the inability to explain acupuncture, strongly signaled the need for a new theory.

One new idea, called the gate-control theory, was first advanced by Ronald Melzack and Patrick Wall in the mid-1960s. They proposed that modulation of pain signals could occur in three ways: First, large fibers in sensory nerves close the "pain gate," while small fibers open the gate; acupuncture needles may work by stimulating large fibers. Second, the pain-signaling system can also be affected by areas in the brain stem; since the brain stem is neurally connected to a large area of the body, stimulation of these brain-stem regions may explain analgesia in those areas. Third, pain is also controlled by fibers descending from the cortex. Far from being a simple switchboard, the body's pain network may more closely resemble one of those huge, computer-operated communications consoles: just as a call from Brooklyn to Manhattan may be more efficiently routed through Detroit, a needle in the knee may be just what the doctor ordered for a pain in the pelvis.

Even if the gate-control theory proves correct, scientists are still going to search for the observable "something" that acupuncture needles stimulate. Recent work at the National Institute of Mental Health has

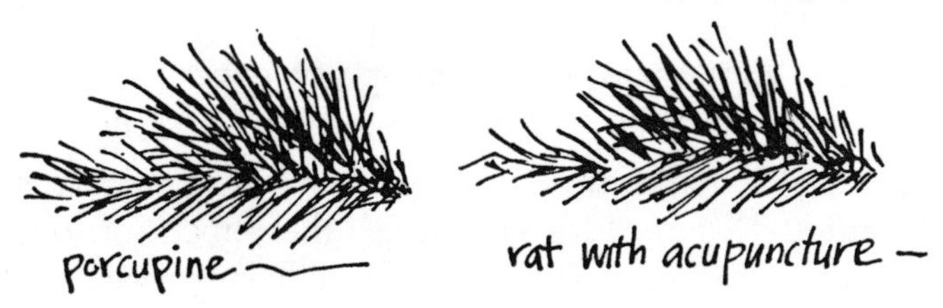

Acupuncture—A Lesson in Animal Identification

uncovered clear evidence that acupuncture stimulates the brain to produce endorphins—those naturally produced chemicals believed to be involved in combating pain. If the acupuncture-endorphin link is confirmed by further experimentation, the AMA may have to capitulate: basic biochemical explanations are the Holy Grail of Western medicine.

One other possible mechanism is the recent discovery of electromagnetic channels surrounding normal nerve cells. Dr. Robert Becker, in his work on regeneration of limbs, has discovered that half the traditional acupuncture points correspond to spots in the nervous system that seem to be amplifiers of electricity. These spots may be "booster stations" for a second network of nerve transmissions; in that case, acupuncture might involve a short-circuiting of these amplifiers to shunt pain signals away from the brain. Work is proceeding with electro-acupuncture and with an even more radical application of technology to folk medicine, laser acupuncture, in which penetrating beams of light are used to replace the traditional silver needles.

CONVERSATIONAL TACTICS:

The following demonstration of acupressure—acupuncture without needles—is strictly X-rated. It's known as "the serpent of love," and reportedly can arouse a hibernating coil of sexual energy within minutes of treatment!

The special trigger points are located in the hollows of each buttock. Insert the tip of your thumb or a knuckle into the flesh in the center of the hollow and massage in a deep, rolling manner for several minutes. Make the movements of your thumb describe slow circles, about three quarters of an inch in diameter. Your partner should first experience a relaxation of the lower abdomen, then a release of muscular tension, and finally, "the serpent's stirring." If you can accomplish all that with an uneducated thumb, just imagine what miracles a trained acupressurist can perform using ten highly trained digits.

RELATED TOPICS:

Appropriate Technology, Consciousness, Endorphins, Placebo, Regeneration.

FURTHER READING:

Acupuncture, Marc Duke, Pyramid Books, 1972.

"How Acupuncture Works," Ronald Melzack, *Psychology Today,* June, 1973.

"Acupuncture for Analgesia and Acupuncture for Everything Else," Robert J. Trotter, *Science News,* October 27, 1979.

ALTRUISM/SELFISH GENE

DEFINITION:

A hypothesis of sociobiology theory which holds that self-sacrificing behavior without hope of reciprocation (altruism) may be genetically encoded in individuals in order to increase the survival probability of the kinship group which shares that gene complex.

WHAT IT REALLY MEANS:

In our great-grandparents' time, Charles Darwin's theory of evolution came as a nasty jolt to the collective human ego. Back in the last century, people liked to think of themselves as candidates for sainthood, just one step below the angels. Naturally, Darwin's picture of our species as a bunch of bald apes with talented thumbs was not exactly what the body politic wanted to hear. The theory was intensely debated in the parlors of Victorian England and in the Tennessee courthouse where a schoolteacher named Scopes was legally tried and convicted of teaching this subversive doctrine.

The echoes of that century-old debate are still to be heard in modern America. If you have any doubt about that, just visit your local PTA the next time they discuss which biology textbook to use. In fact, you're likely to hear the same slogans that were shouted at Darwin and Scopes: "Our kindness toward our fellows is proof enough of our unique divine origin; the golden rule is evidence that man did not descend from the apes." Altruism is a slap in the face of natural selection, say Darwin's critics. Our concern for other humans, even in the face of personal danger, is what makes us far more than a flock of featherless bipeds.

Could even such an admirable human trait as altruism be based on the eons-old molecular structure of our chromosomes, rather than the relatively new inventions of culture and consciousness? Is it possible that altruism is really just a form of genetic selfishness? These questions, and the perspective they represent, have been the principal tenet and one of the chief drawing cards of the new field of sociobiology—a radical theory which claims that our social behavior as well as our physical characteristics are at least partially determined by biological evolution.

According to Darwin's theory, traits evolve over time through the processes of random mutation and natural selection. The genes for big brains, fleet feet, or sharp tusks confer survival value on the organisms who possess them, thus increasing the chances that the fortunate creatures will reproduce and pass the genes to successive generations. Those individuals who are better adapted to the environment survive and reproduce; losers die out.

The theory of natural selection through individual survival and reproductive success—what we commonly call "survival of the fittest"—does, however, appear to have a serious flaw. Why should a man or a woman, or a bird or a bee, jeopardize personal survival to benefit others? If altruism decreases the probability of having descendants, shouldn't the altruism gene be selected out and disappear from the gene pool? After all, the tendency to sacrifice your own place in line is spectacularly unsuccessful, reproductively speaking: it is notoriously difficult to reproduce when you're dead. Yet humans continue to risk their lives for their fellows. Bees die for the hive. It certainly seems that Darwin left something out when it comes to altruism.

An early attempt to save the theory was the doctrine of group selection, which held that the genetic combination defining the tendency for self-sacrifice is an evolutionary survivor because it improves the overall fitness of the group: "My species, right or wrong." Unfortunately, group selection failed to specify what constitutes a relevant group. In the mid-1960s, though, some insect biologists (of all people) developed a theory with greater explanatory power. Entomologist William Hamilton II, working with bees, developed the notion of *kin selection*—an idea that threatens to move all "higher faculties" such as altruism firmly into the realm of self-interest and the survival instinct.

If one looks at the evolutionary contest in terms of genes rather than individuals, there is one way that self-sacrifice can confer evolutionary advantage—if the loss of one life helps to ensure the survival of relatives who share the same genes as the dead altruist. If the sacrifice of one member of a group increases the survival chances of many genetic relatives in that group, the transaction is an excellent bargain in the Darwinian sense. It also makes altruism look more like a maximization of "kinship profit" than a spiritual gift or a creation of civilization.

The furor over the sociobiological interpretation of altruism was captured in the title of Richard Dawkins's book, *The Selfish Gene.* When we look at evolution from the point of view of our genes, it becomes the history of populations of near-immortal organisms composed of DNA. When mortal host organisms reproduce, we modulate the DNA message and pass it down through time. You and I and all our ancestors can be seen as a series of temporary mobile homes for the self-sculpting DNA drama. The genes direct our behavior in ways that maximize their own chance for survival. If deceit, selfish altruism, and subtle forms of cheating happen to confer survival benefit, the genes make sure those traits survive.

CONVERSATIONAL TACTICS:

Discussions about the validity of the selfish gene theory have a tendency to heat up fast. After all, we are not so far removed from the

Altruism/Selfish Gene—Lifeboat Ethics

Victorian era that we enjoy being told that our most human emotions and aspirations can be explained by the same mechanisms that govern the behavior of ants. The human invention of language is indeed a mighty tool, and the primacy of chromosomes over consciousness is far from proven. Would anyone embrace, without solid proof, Dawkins's characterization of the selfish gene in action? "They swarm in huge colonies, safe inside gigantic lumbering robots, sealed off from the outside world, manipulating it by remote control. . . . We are their survival machines."

Not only are the consequences of this theory controversial, so is the evidence. Since no one has ever seen a selfish gene, or specified the molecular code for anything more complex than a specific protein, all evidence thus far is circumstantial and therefore inconclusive and open to differing interpretations. In the case of human altruism, the other interpretation is the phenomenon of human cultural adaptation: People learn new behaviors in response to specific situations and pass on those successful behaviors through the media of science, literature, music, and art, counterbalancing the predetermined messages passed through the chromosomes.

Talking Tech presents no case for either side of the debate. We offer instead the following litmus test of belief in the selfish gene theory, a conversational tactic which is especially appropriate at family reunions. Ask yourself if you would give your life to save your brother, one grandchild, and three cousins. If your answer is yes, you are a believer, because you would save 9/8 of your genetic material (4/8 brother, 2/8 grandchild, 3/8 cousins) by sacrificing yourself (8/8). If you answer no, your genes may not be programmed for kin selection—or you may not be particularly fond of your cousins.

RELATED TOPICS:

DNA, Human Origins, Mutation, Recombinant DNA, Sociobiology.

FURTHER READING:

The Selfish Gene, Richard Dawkins, Oxford University Press, 1976.
On Human Nature, Edward O. Wilson, Harvard University Press, 1978.

ANALOG COMPUTER

DEFINITION:

A mechanical aid to computation in which mathematical variables are represented as continuously changing physical quantities. The physical form of the analog may model the system being studied, or the analogy may be based solely upon the mathematical equivalence of physical behavior and computer-stored variables.

WHAT IT REALLY MEANS:

Look closely at your watch. Are Mickey's hands pointing at the one and the three? If you're wearing any watch with an old-fashioned predigital face—the kind with a big hand and a little hand—then you are already familiar with an analog computing device. Your watch qualifies as an analog computer because a continuously changing physical quantity (the action of the gears, springs, and hands) represents a mathematical variable (the time).

In order to master the vocabulary of analog computation, you need to learn only three words: *model, measure,* and *continuous.* Analog machines are constructed to be physical *models* of the problems they are to solve: The inner workings of a watch model the spinning of the earth. Because physical quantities are used, analog output is determined by *measuring* changes in these quantities. The distance the watch gears move is measured to compute the time. The output of an analog device is *continuous* rather than discrete. The watch hands are always moving; when we say it is 3:05, we have "frozen" their motion to give an accurate approximation of the time.

When it comes to building a truly powerful, versatile computer—the kind that messes up your bank balance and makes your airline reservations—analog machines are not as useful as the *digital* kind. Digital computation breaks information into discrete units instead of using a smooth, continuous flow. *Digitus* is the Latin word for fingers: counting on your fingers is the quintessential digital activity. Sundials and sextants are analog. Pocket calculators and televisions are digital. Except for limited industrial uses, analog computers are the dinosaurs of the computational world, and digital devices are the new, dominant species of the solid-state cosmos.

CONVERSATIONAL TACTICS:

Before the era of the cheap pocket calculator, every science and engineering student's rite of passage included initiation into the mysterious operations of the slide rule—the most popular analog computer of all time. Now, in less than ten years, these cunning devices have fallen from the position of powerful scientific tools to that of historical curiosities. In the spirit of sci-tech nostalgia, *Talking Tech* presents directions for constructing and using your very own slide rule. Who knows? Maybe you'll be stuck without batteries sometime when you need to make an important calculation.

Trace or photocopy the two scales on this page and mount each one on a heavy piece of cardboard. If you trace them, you must do it exactly—the markings are accurate to 1/32 inch:

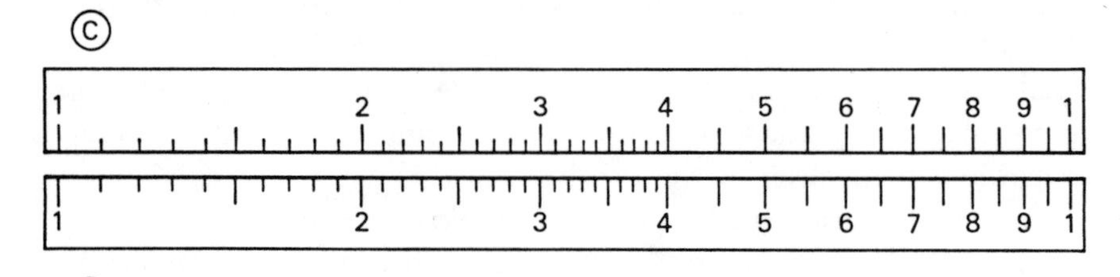

Analog Computer—An Ancient Computer

The key to the arcane art of slide ruling is remembering that everything on the slide rule is a power of ten, so it is up to you to decide where to put the decimal point: 2 × 2 could be 4, or 400 (20 × 20), or 40 (2 × 20), or 4000 (200 × 20). By now you're beginning to see why a slide rule is considered to be an initiatory ordeal.

Try a multiplication example: thirty-three × twenty-seven. Move the left side's 1 on the D scale directly under the 33 on the C scale and read the mark across from 27 on the D scale—almost 9. You know that 30 × 30 equals 900, and that since 3 × 7 is 21, your answer must end with a 1, so you deduce that 891 is your answer. If it seems like a hopelessly complicated system, remember that engineers using slide rules devised the digital calculators everyone uses today. Slide rules work, but they take a little practice before you get used to them.

RELATED TOPICS:

Digital Computer, Information, Mathematical Modeling, Miniaturization.

FURTHER READING:

Logic, Computing Machines, and Automation, Alice Mary Hilton, World, 1963.
Connections, James Burke, Little, Brown, 1978.

APPROPRIATE TECHNOLOGY

DEFINITION:

Because appropriate technology (AT) is as much a social, political, and economic concept as it is a scientific and technological one, there is no single, precise definition. Attempts at elucidation run from cryptic slogans like "technology with a human face" to complex definitions relating AT to economic development.

WHAT IT REALLY MEANS:

The real meaning of appropriate technology seems to lie in the mind of the beholder. As with alternative energy, one has to ask, "Appropriate to whom?" and "Appropriate in what respects?" The thinking wasn't always so murky. AT started out in the 1960s as the perfectly respectable economic concept of "intermediate technology" (IT), which E. F. Schumacher defined as "more productive than the indigenous technology but immensely cheaper than the sophisticated, highly capital intensive technology of modern industry." He then listed four characteristics that any such technology should exhibit: (1) it should create rural employment; (2) it should be labor-intensive rather than capital-intensive; (3) the technology should not require a skilled labor force, and (4) production should be for local use. While it was easy enough to have a legitimate dispute about whether or not intermediate technology was viable as a theory of economic development, the concept at least seemed well-defined.

Unfortunately, IT had a crucial semantic flaw. No individual or nation wanted to settle for "intermediate" anything. Even if it was just a stage before the attainment of high technology, "intermediate" sounded like settling for less, and it had to go.

"Appropriate" took its place, and then a peculiar thing happened—peculiar for the normally obscure world of theoretical economics. When Schumacher published *Small Is Beautiful* in 1973, it immediately struck a responsive chord. Citizens of highly developed nations, most notably the United States, were already starting to rebel at the perceived inability of high technology to solve their problems and the tendency of centralized technology to control their lives. Somehow, even though it was never adequately defined, AT seemed like the answer. How could a concept which had its strongest impact on developing nations be translated into the context of U.S. society? The spinning wheel is a great advance for an African woman who uses a drop spindle; however, it can hardly compete with huge automated textile mills.

Rather than becoming a scientific theory, AT became a series of slogans: "The purpose of AT is to make people more self-reliant"; "Production by the masses, not mass production"; "Any technology gradually creates a totally nonhuman environment." Depending on whom you were talking to, AT could mean appropriate to economic objectives (decentralization) or environmental objectives (cleaner air and water) or energy objectives (self-reliance). By the mid-1970s, AT was a movement and no longer just an economic theory. Governor Brown of California became one of AT's most influential converts for a while, and the growing antinuclear movement also helped usher AT into the political arena.

Of course, lack of a generally agreed-upon definition doesn't mean that the family of concepts which coalesce under the AT label are deficient or not useful. The general backlash to the "bigger is better" and "technology can solve all our problems" dogmas, as evidenced by the consumer movement, environmental movement, and AT have all made contributions to the quality of our lives. Solar home heating is an appropriate technology. So are compost heaps, home vegetable gardens, and bicycles. However, in a certain sense, so is the Hoover Dam, fertilizer factories, and automobiles. The right technology needed to do the job is the appropriate technology.

CONVERSATIONAL TACTICS:

Appropriate technology is bound to be a topic of conversation for some time to come. Like all movements, AT has developed a jargon of its own. In order to prepare you for a discussion of AT, *Talking Tech* presents this glossary of AT buzzwords:

charka—the type of spinning wheel used by Gandhi; the first AT device and virtual Holy Grail of the movement.

Club of Rome—a group of international economists who periodically publish reports (e.g., *The Limits to Growth)* portending global doom if industrial and population growth are not curtailed.

The Whole Earth Catalog—the Sears Roebuck Catalog of the AT sphere.

Coevolution Quarterly—a spin-off from *The Whole Earth Catalog*; one of the most popular AT journals, followed by *Mother Earth News.*

Green Revolution—originally, the development of high-yield crops; now used as a more general reference to a more rural "back-to-the-land" life-style.

soft path—a term used by Amory Lovins to describe an energy technology which relies on renewable resources and a decentralization of energy supply. The goal is to match scale and energy availability with end-use requirements.

walking on two legs—the Chinese technological model, reliant on both small, labor-intensive technologies and large-scale, capital-intensive techniques.

RELATED TOPICS:

Acupuncture, Energy Alternatives, Meltdown, Technology Assessment, Weather Modification.

FURTHER READING:

"Modern Technology: Problem or Opportunity," *Daedalus,* Winter 1980.

Paper Heroes: A Review of Appropriate Technology, Witold Rybczynski, Anchor Books, 1980.

Small Is Beautiful, E. F. Schumacher, Harper & Row, 1973.

ARTIFICIAL INTELLIGENCE

DEFINITION:

The development of machines which can perform tasks that would require intelligence if done by human beings.

WHAT IT REALLY MEANS:

Artificial intelligence (AI) has only just emerged from the laboratories and entered our lives, but developments in the prototype stages are proceeding rapidly. Robots on Mars, under the sea, and in the cores of nuclear reactors are already performing vital tasks in environments inhospitable to human intelligence. Existing data bases and computer programs can now administer psychological tests, design molecules, conduct legal research, and help diagnose disease. Whether or not these machines "think" and are "intelligent" is as much a question for philosophers as for technologists.

In order to understand what AI is, it helps to clarify what it is *not.* First you must recognize what may seem like a purely semantic distinction: AI is not a computer, a robot, or any other kind of machine. AI is something that computers and robots may someday be able to *do.* It is a *capability* rather than a mechanism.

Next, forget all those myths about robots taking over our lives, because they have already done that. Our power and communication grids, transportation systems, defense apparatus, cities, hospitals, economic institutions, and industries have been controlled by computers for years, and our lives would be in a chaotic state indeed if all those artificial, high-speed, interlinked, general-purpose symbol-manipulators were to dissolve tomorrow morning. Much of the hardware for tomorrow's world-brain is in place *today;* in that sense, we all live within the prototype of an AI. In 1969, when a billion people watched a live television transmission from the moon, they became part of the first tentative thoughts of the growing planetary nervous system.

Another good thing to remember while considering the philosophical problems of "artificial" intelligence is the evidence that our own brains evolved into "natural" intelligence because our prehistoric forebears were stimulated by their use of tools. Some naked apes learned to stand on their hind legs, freeing their hands to pick up things and throw them. A hundred thousand generations later, the balancing-and-ballistic computer inside the human skull had gained over a pound of potent gray matter.

The overall game hasn't changed as fast as our brains. The first truly portable electronic computer was designed to fit in the warhead of a missile, so intelligence is still connected to our predilection for throwing things at each other. For better or worse, tools and human minds have been shaping each other since our distant ancestors left the trees. In the AI laboratories, the long brain-shaping/tool-building game approaches its logical conclusion.

When digital computers were invented during World War II (in order to solve ballistics and other war-related problems), an unlikely alliance of philosophers, physicists, logicians, psychologists, and mathematicians started to consider seriously the possibility of using these new calculating devices to eventually construct an artificial intelligence. The very first problem—how to define the criteria for recognizing a true thinking machine—was given a tentative solution in 1954 by one of the great pioneers in the field, A. M. Turing, who devised a simple riddle that has come to be known as the *Turing test:* If a human questions a device via a teletype and is unable to judge, solely on the basis of the answers, whether a machine or another human is answering the questions, then the machine possesses true "adaptive intelligence."

The approaches to AI are as complex and varied as the goal itself. First, there is the distinction between the theoretical and experimental branches of AI research. One school concentrates on learning what thinking is, by studying people as well as machines—how can you build a device to do a task the designer does not understand? The other branch skips the psychologizing and gets right down to cases—how to build artificial perception devices, robots, etc. Both theoreticians and engineers converge on AI from five directions: game playing, perception, problem solving, robotics, and speech-language.

AI proponents take their game playing seriously, because activities such as chess, checkers, and backgammon provide important models of human problem solving. Although the domain of the problem is restricted to the board, machines are rapidly proving to be more than a match for human competition. Computers became unbeatable in checkers several years ago, backgammon champion Paul Magriel was recently blitzed in a head-to-"head" match, and only grand masters can beat the best of the chess programs. Current AI work is developing programs for games which are not board-dependent. Poker, a game which requires a lot more than simple logic, is now computerized—and tests have shown that human players cannot differentiate the machine player from the other human players.

Perception, specifically the ability to correctly discern auditory and visual stimuli, is an absolutely vital human skill, and one that AI has long sought to mimic. It took years simply to program a computer to recognize the letters of the alphabet. Today, machines not only know their alphabet, they can read aloud: the Kurzweil Reading Machine can now scan a page, recognize words and letters, apply phonetic rules, distinguish phrase boundaries (cues to how humans split up words and passages in speech) from syntactical and vocabulary knowledge, figure out where stress and accent should go, and finally synthesize it into the "spoken" word. The talking appliance is an unavoidable and imminent development.

Impressive as the Kurzweil device may be, perception at this level is restricted in the same sense that game playing is a board-constrained version of human problem solving—the Kurzweil visual universe is totally described by twenty-six letters. AI is now working to enable computers to process visual information in quasi-human fashion, using heuristic programming in combination with image-enhancement technology. Marvin Minsky, head of the AI laboratory at MIT, thinks that our brains store a number of handy reference pictures. This means that each perceptual experience doesn't have to start from scratch; rather, new visual information is compared and contrasted to that data already on file. How that is done, and what is on file, are major issues of cognitive psychology as well as AI.

Perhaps the essence of being a human is the ability to be a general problem solver. No one quite knows how we do this, but evidence indicates that it's a heuristic process rather than an algorithmic one. *Heuristics* are rules of thumb, general procedures which we follow without guarantee of success. *Algorithms* are precise sets of rules which enable us to solve specific problems. As with other branches of AI, scientists have achieved their greatest problem-solving successes in circumscribed domains. Thoroughly describe a universe of different-colored and -shaped blocks to a problem-solving robot, and it will devise a strategy for retrieving a specific block from a range of possibilities which might include stacking the blocks, using one block to push another, or using one as a tool to reach the target. This is a much more sophisticated means of solving the problem than simply programming the computer to identify block A and block B, stack block B on block A, move both blocks exactly fifteen centimeters, etc. Admittedly, this is the type of problem-solving behavior in which chimpanzees engage, yet it is a breakthrough in heuristic programming—the avenue most AI specialists see as the best route to producing general problem-solving machines.

Robotics, that branch of AI research devoted to building devices which can manipulate objects and move from place to place, has had a glamorous fictional history which has been well in advance of the actual state of the art. From the Tin Man in *The Wizard of Oz* to C-3PO of *Star Wars,* the media have glorified the notion of a metal, mechanical man. However, more often than not, robots of recent vintage have been closer to R2-D2 than to the Tin Man. One of the earliest mobile robots, developed at Johns Hopkins and known as "the Hopkins Beast," resembled a garbage can on wheels, SRI's "Shakey," its name a comment on its lack of surefootedness, looked like a desk-top copy machine perched on a mobile base, and other vintage robots have resembled vacuum cleaners and huge cooking pots.

Today, however, robot research is less concerned with building a general-purpose mechanical man and more concerned with specific industrial applications. Robotics has even changed its name. *Telepresence* is the new science of high-quality, sensory-feedback, remote-control tools. While we're still far removed from having tools which can manipulate with the skill of the human hand, some primitive telepresence devices are currently being used. The Alvin submersible, a kind of mini-submarine, has appendages which can reach out and pick up samples. Much of our space program, the *Viking* Mars missions, for example, depends on remote-control devices to manipulate alien environments. Closer to home, General Motors is using vision-equipped manipulators called "hand-eye machines," and Kawasaki is planning factories entirely run by robots. In nineteenth-century England, the Luddites started wrecking factories because of steam power. How will twenty-first century people respond in the face of telepresence?

One of the first great hopes of AI was machine use of natural language. Things didn't go so well at the beginning, and computerniks came in for a lot of criticism. One of the stories, probably apocryphal, that went around at the time was about the machine that translated the

Biblical verse "The spirit is willing, but the flesh is weak" from English to Russian and back again. The final wording: "The wine is agreeable, but the meat has spoiled." It still seems unlikely that a universal translator for all types of speech can be built, but AI has made some impressive strides in the last thirty years.

You can now buy a hand-held language translator that translates and pronounces words (in English, French, German, and Spanish) punched in through a keyboard. A computer with the ability to listen to human speech grants employees of the state of Illinois clearance to make long-distance phone calls. A computer program named PARRY, designed to respond like a paranoid, was sophisticated enough to deceive six psychiatrists who couldn't tell PARRY from a real paranoid. Another program, named ELIZA, can actually simulate a nondirective psychotherapist by reflecting the patient's statements. You may never want to talk to a psychotic computer or a computer psychiatrist, but the odds are high that sometime in the future you'll end up talking to a machine.

If you think having a multilingual, chess-playing robot clanking around your home sounds like science fiction, consider the even more imaginative possibility that you could increase your brain power by allying yourself with that friendly AI device. Such *human intelligence amplification* research seeks to put the human nervous system in direct contact with computer circuitry in order to form a perfect interface between us and machines. Although this sounds like a familiar science fiction plot, the technology for brain interfacing is currently under development.

The magnetoencephalogram (MEG) is a highly sensitive device for measuring the magnetic fields generated by neural activity. Since these fields are represented by electromagnetic signals, they can be directly entered into the computer. Programming in the future will have nothing to do with punch cards and artificial languages, because we'll probably don our MEG caps and address the computer directly through our thought processes. Both artificial and biological intelligences may benefit from such an arrangement: human brains can store more information and process it in more novel, creative ways, while computers can process information millions of times faster. It might not be such a frightening prospect to become an extension of a computer if it also becomes an extension of us.

CONVERSATIONAL TACTICS:

The most heated topic of Victorian parlor debate was the theory of evolution. Had we really descended from the apes? It is possible that one of the hottest topics for debate in the next century will be the theory of computer evolution. Did computers once evolve from the lowly minds of *Homo sapiens?* Whatever they're talking about a century from now, you can be sure that the next twenty years will see plenty of lively discussion regarding the question: Can machines think? You should be prepared.

For those who wish to argue the affirmative—that machines already can think, or will in the future—the tactic is quite clear. Computers are only in their infancy, but they are evolving rapidly. The experience of the last three decades is that computer capacity increases by a factor of 10 every seven to eight years. At this rate, computers produced in the early twenty-first century will have roughly the same memory capacity, and some of the thinking capability, of humans. This argument is simply one of empirical progress: assuming that there is no inherent reason why machines can't think, we are making good progress toward the goal, so it is only a matter of time.

Arguing the negative—that machines cannot now and never will think—is much more difficult. It requires what is known in the trade as an impossibility argument: No matter what technological advances are made, or could be made, there is something inherent in the concept of machines which makes it impossible for them to think. Such arguments are notoriously difficult to prove, but in the heat of a debate they may prove to be just the right conversational ammunition.

Here are a number of such arguments which other thinkers have advanced:

The theological objection: Thinking is a power God has bestowed on humans. No inanimate machine will ever be able to think.

The consciousness objection: Thinking implies feeling and emotions. The essence of thinking is knowing that you're thinking, and this implies consciousness. Since machines can never be conscious, they can never think.

Lady Lovelace's objection: Lady Lovelace was a nineteenth-century English lady who reacted negatively to the idea that Charles Babbage's analytical engine (an early computer) could ever think. The argument is that computers can't originate anything—they can only do whatever we know how to order them to perform. Since computers can't create, they can't think.

The nervous system objection: Since the nervous system is not a discrete state machine, as are digital computers, such machines will never be able to think, since that requires the type of mechanism found in our nervous system.

The body objection: It is not just with our brains that we think; our bodies and the way they interact with our brains produce thought. Since digital computers are brains without bodies, they will never think.

RELATED TOPICS:

Consciousness, Digital Computer, Gödel's Incompleteness Theorem, Intelligence Enhancement, Mathematical Modeling.

FURTHER READING:

Gödel, Escher, Bach: An Eternal Golden Braid, Douglas R. Hofstadter, Basic Books, 1979.

What Computers Can't Do: A Critique of Artificial Reason, Hubert Dreyfus, Harper & Row, 1980.

Machines Who Think: A Personal Inquiry into the History and Prospects of Artificial Intelligence, Pamela McCorduck, W. H. Freeman, 1980.

BEHAVIOR MODIFICATION

DEFINITION:

This is a technique for manipulating personal conduct based on behaviorist theories. Without regard for "internal" factors or motivation, these theories maintain that it is possible to modify the conduct of humans and animals by identifying the environmental rewards and punishments which control behavior and arranging appropriate reinforcement schedules.

WHAT IT REALLY MEANS:

Although the implications of his theories are still hotly contested within and beyond the world of psychology, no one can deny that B. F. Skinner has played a singular role in the development of the science of human behavior. For thousands of years, natural philosophers tried to explain "what makes people tick" in all sorts of peculiar ways. This led to the pseudotheories of "humours," demonic "possession," and assorted "vital essences." At the turn of the twentieth century, Freud, Jung, and their psychoanalytic colleagues took the search deep inside the self, formulating the concepts of *id, ego,* and *superego* to explain what they found in the subconscious minds of their patients.

By the 1940s, Skinner and the other behaviorists decided they had heard enough of what they considered to be idle speculation. These psychologists began by declaring that too little was known about *how* organisms behave to formulate any useful theories about *why* they do what they do. Instead, they focused on the visible output of the mind—behavior. By concentrating on conduct—what persons or pigeons *do,* instead of what they *think*—and by emphasizing prediction and manipulation over explanation, the behaviorists were able to develop powerful techniques for shaping behavior.

Operant conditioning is the technique that makes behavior modification possible. In contrast to Pavlov's "conditioned responses" made directly in response to stimulation, Skinner focused on responses emitted by the organism in the absence of any apparent external stimulation. These responses were then positively or negatively reinforced as soon as they were emitted. Quite simply, operant conditioning relies on increasing the frequency of targeted behaviors by rewarding them after they appear spontaneously.

At the common sense level, operant conditioning does furnish a handle on human behavior—everyone knows that people seek pleasure and avoid pain. If you bump into something and you find it acutely painful, you will be less likely to bump into it in the future. If a rat accidentally pushes a lever and a pellet of food then drops into its dish, it tends to push the lever more often. If the same rat is given an unavoidable shock whenever a minute passes without it pressing the lever, the rat ends up pressing the lever at least once a minute.

Although the generalized connections between the way a rat in a cage behaves and the way a human being behaves in society are cause for argument, few people can deny that behavior modification works very well on people as well as rats, and it doesn't seem to matter whether or not the people know that their behavior is being shaped.

Through the careful collection of data from animals and humans in controlled environments, the behaviorists first formulated their "functional analysis of behavior": A first step toward behavior modification is to devise a system for identifying target behaviors and isolating the environmental factors controlling those behaviors. In the case of laboratory animals, and in the case of people to some extent, a few potent primary forces provide that control—hunger, thirst, sexual desire, the need for sensory stimulation. Any stimulus which can relieve deprivation of one of those needs is a *reinforcer,* the basic building block of behavior modification.

Especially important are *schedules* of reinforcement: Do you reinforce the subject every time it exhibits the target behavior? Do you reinforce a fixed number of responses? Do you only reinforce it once in a while? This last schedule, known as *intermittent reinforcement,* is one of the most potent, probably because it is closer to the way things happen in real life. An example of intermittent reinforcement is the random payoff of a slot machine; visit a casino and watch those rows of humans hypnotized by machines which are set to a precisely determined reinforcement schedule.

Lever-pressing and one-armed-bandit pulling aren't the only modifiable forms of behavior. By using *successive approximations,* it is possible to shape complex behavior. If, for example, you want to teach young children to sit quietly for a minute, you have to start by rewarding them for sitting still for five seconds, and when they learn that, you have to withhold the reward until they sit quietly for ten seconds. Then you can link the reinforcement schedule to a more complex behavior, like sitting quietly and doing homework.

One of the more intriguing findings, which certainly does seem to apply to people as well as pigeons, is that a *random* reinforcement schedule, not contingent on anything the subject does, produces "superstitious behavior." Skinner discovered that if you give a hungry pigeon a piece of grain every ten seconds, no matter what it does, something remarkable happens. If you leave a group of separated subjects in this situation and return to check them several hours later, you will find the pigeon in cage one turning in clockwise circles, the pigeon in cage two bobbing its head compulsively, the specimen in cage three hopping from one leg to the other—an unpredictable range of totally unrelated behaviors. That is where the behaviorist version of "superstition" comes in: Whatever a particular subject happened to be doing when the first reward hit the feeding pan—turning slightly to one side, gently bobbing its head, etc.—it tended to keep on doing. Are people so terribly different from pigeons in cages when it comes to superstitious behavior?

CONVERSATIONAL TACTICS:

One of the great strengths of operant conditioning is that you don't need expensive equipment or extensive training to perform behavior-shaping experiments on your friends and family. All you need is the knowledge of a few simple techniques. Those hapless folks around you don't even have to *know* that you're manipulating their actions! Of course, important ethical questions arise when you modify another person's behavior without their consent (see Bioethics), but just a dab of

Behavior Modification—Superstitious Pigeons

clandestine experimentation can lead to a behavioral-tech conversational coup when you finally reveal what you've been doing.

You need a conversational partner of any size, sex, or age, a twenty-minute experimental period, and the ability to unobtrusively observe the time. (Hint: This environment corresponds closely to the ritual of sharing cups of coffee.) Divide the experiment into two blocks of ten minutes each. For the first ten minutes, reinforce your partner's opinion-stating behavior by strongly *agreeing* with any statement beginning with the words "I think . . ." or "It seems to me . . ." During that same ten-minute period, secretly tally the number of times your partner states an opinion. Then, when the second experimental period begins, respond to opinion statements either by remaining silent or by strongly *dis*agreeing; without revealing what you are doing, tally the number of opinions your partner expresses during the second experimental period.

When your twenty minutes are up, tell your friend that you have just modified his or her behavior. If he disagrees, you have your secret tallies to back your assertion with hard data. It always sounds convincing when you say: "You voiced forty-eight opinions when I was reinforcing them and only twelve opinions during the time I was not reinforcing them." Then you can move on to what you know about operant conditioning, superstitious behavior, successive approximation, etc.—or you can segue to another area and discuss the ethics of clandestine experimentation.

RELATED TOPICS:

Bioethics, Cognitive Dissonance, Consciousness, Pheromones, Placebo.

FURTHER READING:

Systems and Theories in Psychology, M. Marx and W. Hillix, McGraw-Hill, 1963.

Beyond Freedom and Dignity, B. F. Skinner, Knopf, 1971.

BIG BANG

DEFINITION:

A cosmological theory explaining that the universe began approximately fifteen billion years ago when an infinitely hot, infinitely dense singularity exploded. The theory has been supported by laboratory observation of high-energy particle interactions and radio-astronomical isolation of microwave background radiation.

WHAT IT REALLY MEANS:

Creation myths are one of the few truly universal cultural traits. Every tribe, culture, nation, or religious cult in history has concocted some version of how everything got started. Modern societies are no different from previous groups, and the cosmogonists are the creation mythologists of our science-dominated culture. What is different this time is the modern mythmakers' claim that their brand of creation story can be confirmed through objective measures. Strangely enough, those scientific measures and the ancient mythologizers agree on a number of crucial points.

The Book of Genesis, the Bhagavad Gita, and the big bang theory all agree that *fiat lux*—let there be light—is an excellent description of the way it all started.

According to big bang theorists, there was plenty of light in the first microseconds of time—light such as the universe hasn't witnessed since. The Old Testament didn't go into detail about positrons and neutrinos, but it most definitely mentioned light, and contemporary physicists do agree that photons (light particles) were an important component of the universe during the opening scene. However, the light of that primordial dawn was physically different from the diluted light we see today. For one thing, it was a billion times denser than water. Another astonishing characteristic of that flash is that evidence of it has lasted fifteen billion years, which is fortunate for astrophysical theory, since it is the only data earthbound cosmogonists have to work with.

Our idea of the very limits of creation—how and when it began, and how it might end—was vastly expanded by a pair of American astronomers, Edwin Hubble and Milton Humason, who conducted a series of observations in the 1920s. Working with blurry lines of light on old-fashioned photographic plates, these astronomers found evidence that led to our current model of an expanding universe.

If all the galaxies are expanding away from each other at a measurable rate, scientists theorized that it should be possible to trace the current expansion backward to a specific time in the past. This "tracing-it-all-back" process led to the big bang model. According to this theory, our universe is expanding today because it exploded from a single point sometime around fifteen billion years ago. This was the big bang. But it wasn't an explosion in the sense that we understand the term. It wasn't a spherical wave that started at a central point and spread outward into space, like a bomb detonation. It was a total explosion that happened everywhere in space at the same time. Everything that existed rushed away from everything else. In a real sense, the universe *is* that explosion. It was such a singular event that present-day scientists are still not equipped with theories sophisticated enough to describe the events of the first millionths of a second. Current astrophysical tools begin to be useful only when the first one hundredth of a second is reached.

At around one hundredth of a second after the beginning of the bang, the temperature of the universe was around 100,000 million degrees Kelvin—a temperature that makes the interior of an exploding star seem very cool in comparison. Today, however, that incredible blast of ultradense radiation and colliding particles has thinned out and cooled down to an average cosmic temperature of about three to four degrees Kelvin. While this temperature is not the kind that can be measured by sticking a thermometer into space, the big bang theorists postulated that residual heat should still be detectable as radiation in the microwave range.

In 1965, A. A. Penzias and R. W. Wilson at the Bell Telephone Laboratories in New Jersey were building a radio telescope and were having trouble with persistent interference in the 7.35 mm frequency. Like good scientists, they tried to eliminate all possible sources of the interference: They scrubbed bird droppings off the antenna, used special supercooled noise-elimination gear, painstakingly recalibrated the antenna. Nothing they did eliminated the interference, and that led to the inescapable conclusion that the noise was coming from *everywhere.* The predicted "fossil radiation" from the big bang had been discovered. It was an awesome confirmation of an ultimately abstract theory; those radioastronomers in New Jersey had actually tuned in on the echo of creation. The big bang theory is now widely accepted as the "standard model," and our view of the cosmos can never be the same again.

CONVERSATIONAL TACTICS:

To travel fifteen billion years back in time, you need only a television set. If you believe that a picture is worth a thousand words, you can punctuate your explanation of cosmic background radiation by actually observing it! Turn on a television set and look at the "snow" on the screen: You will be viewing evidence of the big bang, because one component of that interference is microwave radiation in the 7-mm range. Those random white dots on the screen are caused, in part, by the echo of creation.

RELATED TOPICS:

Black Holes, Cosmology, Galaxy, Radio Astronomy, Red Shift, Stellar Death.

FURTHER READING:

The First Three Minutes, Steven Weinberg, Bantam, 1977.
The Big Bang, Joseph Silk, W. H. Freeman, 1980.

BIOETHICS

DEFINITION:

The systematic study of human conduct and moral values as they relate to the life sciences, health care, and experimentation involving human subjects.

WHAT IT REALLY MEANS:

Philosophy is relevant again. Twenty-five centuries ago, in ancient Greece, the most important questions—from what the world is made of to why people act the way they do—were decided by philosophers. This is no longer the case. Today's "wise men" are the scientists and technologists. The irony of this situation is that it is precisely the success of biology and medicine that has revitalized the dormant discipline of ethics.

Although science originally displaced philosophy as a way of gaining knowledge about the world, the achievements of science are now raising questions that only philosophy can answer: When should a fertilized ovum receive the rights of personhood? To what extent should science alter the biological mechanism of reproduction? When is a person dead?

Who has a right to "extraordinary medical means" as a method of life support, and who has the right to terminate them? What are doctors' responsibilities to their patients?

If these questions sound too abstract, the kind of problems you might expect in a philosophy classroom but not in the real world, consider these names from the headlines of the last few years: Dr. Kenneth Edelin, Louise Brown, Karen Ann Quinlan, Lewis Washkansky, and Dr. Peter Bourne. Dr. Edelin was accused of terminating the life of a viable fetus after an abortion. Questions about contraception, sexuality, and sterilization are currently being studied by bioethicists. Future forms of reproductive technology such as cloning, parthenogenesis, and sex selection promise to provide even thornier dilemmas. Louise Brown owes her very existence to one of these controversial biomedical innovations—she was the first publicly acknowledged "test-tube baby." What will be the name of the first human clone, and will the world ever be the same after that?

Karen Ann Quinlan was the young New Jersey woman who lapsed into a vegetative state, and whose parents sought permission to turn off her life-support equipment. An absolutely vital question in this era of technomedicine concerns the criteria of patient death and the consideration of the patient's right to die with dignity. Definition and determination of death, euthanasia, and ethical suicide are other metaphysical by-products of medical technology.

Lewis Washkansky was the first heart transplant recipient. As bioengineering advances to the point of wide-scale application of artificial, transplanted, or cloned organs, these combinations of human, machine, and surgical wizardry will raise new questions: Who can assign priorities to scarce organs? Is it ethical to keep a brain-dead cadaver on life-support systems as a kind of "organ farm"? What are the ramifications of a demographically aging society? What kind of social chaos is going to break loose if a life-span-increasing drug is discovered?

Dr. Peter Bourne was President Carter's medical adviser who was forced to resign after it was disclosed that he wrote a drug prescription for a phony patient. Medical ethics, the study of value-related problems which arise in the physician-patient relationship, is a major subset of bioethics: Is it ethical for a doctor to lie to a patient? How much does a patient have a right to know about his condition?

The above-mentioned five areas hardly encompass the entire scope of bioethics. In fact, the recently published *Encyclopedia of Bioethics* goes on for two thousand pages about the subject. Some of the other hot bioethical debates concern genetic engineering, behavior modification and control, environmental ethics (what are the rights of future generations?), biological weapons, and medical malpractice. If you still think bioethics does not directly affect you, consider the case of human experimentation.

Twenty years ago, the CIA secretly dosed unsuspecting citizens with powerful psychedelic drugs; several of these "experiments" resulted in insanity or suicide. Thirty years ago, Army researchers released various bacteria in the New York City subways and sprayed them into the air in San Francisco Bay to clandestinely study how biological warfare could be used against the population of an urban center; at least one documented fatality has come to light. Officially, these abuses of public trust and safety don't occur anymore, at least on such a large scale, because of the principle of *informed consent*—researchers must have your permission prior to experimenting, and that consent must be intelligently given, based on a full understanding of the inherent risks. In 1974, Congress created the National Commission for the Protection of Human Subjects of Biomedical and Behavioral Research and charged it with devising guidelines regarding human experimentation.

CONVERSATIONAL TACTICS:

The kind of questions studied by bioethicists—life and death, environmental control, behavior control—are too important to the general population to be limited to academic seminars. There are real people out there with real problems, such as turning off a relative's respirator after brain death, or fixing responsibility for medical bills associated with toxic chemical spills. Needless to say, questions like these are beginning to appear in courtrooms. To help you keep up with the latest trends in bio-law, *Talking Tech* provides you with these test examples, based on real cases in U.S. courts.

Example A:

Mr. Jones, eighty-three, is in a coma; EEG readings indicate that his brain is irretrievably damaged, but he is kept breathing by a respirator. Despite his family's wishes, the doctors refuse to pull the plug; the family sues to stop extraordinary medical treatment. If you were on the jury, would you decide:

a. Case dismissed, because medical care is a matter for doctors, not courts.

b. The court will authorize turning off the respirator if the doctors decide that Mr. Jones can never regain brain function.

c. It is the duty of every physician to keep every patient alive as long as possible by whatever means necessary, and failure to do so is homicide.

(ANSWER: Your judgment is affirmed if you decided b. Upon neurological proof of brain death, the court ordered that the plug be pulled, and Jones died.)

Example B:

A distinguished physician removed a human egg from a woman, put it in a glass flask, and fertilized it there with her husband's sperm. The physician's superior, opposed to the nature of the research, aborted the experiment by destroying the contents of the flask. The parents sued the physician's superior for destroying their potential baby. If you were on the jury, would you:

a. Award the parents damages on the grounds that to them the contents of the test tube were a great deal more than just a laboratory specimen.

b. Dismiss the case because the experiment was unethical and the superior had the right to terminate it.

c. Find for the physician's superior, affirming his medical opinion over the emotional but medically inexpert judgment of the parents.

(ANSWER: In the late 1970s, a Florida couple was awarded fifty thousand dollars in damages in this case, so if you chose a., your bio-legal judgment is correct—until a higher court changes everything.)

Example C:

The state brought suit against the parents of an adolescent suffering from Down's syndrome (mongolism), urging that the child be allowed to have an operation to repair a congenital heart defect. With the operation, the child would probably live a normal life span; without it, he was assured of dying within a few years. The parents argued *against* the operation on two grounds: first, there was only a 95 percent chance of success; second, if the operation succeeded, the child would outlive his parents and become a burden on the state or his siblings. If you were the judge, how would you decide?

a. With the parents, against the operation.
b. With the state, for the operation.
c. Let the child decide.
(ANSWER: a.)

RELATED TOPICS:

Behavior Modification, Clone, Consciousness, Longevity—Immortality, Placebo, Recombinant DNA, Reproductive Technology, Technology Assessment, Toxic Chemicals, Weather Modification.

FURTHER READING:

The Biocrats, Gerald Leach, McGraw-Hill, 1970.
Lying: Moral Choice in Public and Private Life, Sissela Bok, Vintage, 1978.
Encyclopedia of Bioethics, Warren T. Reich, Editor-in-Chief, The Free Press, 1978.

BLACKBODY RADIATION

DEFINITION:

$$I(\omega) = \frac{\omega^2 kT}{\pi^2 c^2}$$

This equation relates the wavelengths and temperatures of the thermal radiation emitted by a theoretically perfect absorber of all incident radiation.

WHAT IT REALLY MEANS:

The search for an explanation of blackbody radiation was a crucial event in the history of physics. This simple laboratory problem led to the revision of classical mechanics and the beginning of the quantum revolution. Physicists started out trying to understand a small inconsistency in a specific experiment and ended up totally redefining the nature of reality. When Max Planck rejected the nineteenth-century wave theory of radiation, he opened the door to all the promethean surprises of twentieth-century physics.

At the end of the last century, the simple wave picture of radiation was beginning to be questioned. It was already known that electromagnetic waves formed a continuous spectrum that stretched from radio waves at the lowest frequency, through visible light in the middle, to gamma rays at the highest frequency. With that knowledge, the question arose: Given a "perfect radiator" or "blackbody" (so called because a body which could absorb and emit radiation with 100 percent efficiency would look black), is there a definite relationship between the amounts of radiation of different frequencies taken in and given out at different temperatures? It was a reasonable question for a physicist to ask. Yet the answer was so unexpected and shocking that it pulverized the foundations of the old Newtonian world model.

Classical mechanics answered the blackbody question in two ways. First, there must be a definite balance at any given temperature between the waves of different frequencies. Secondly, while the perfect radiator gives off waves of all kinds, the waves should predominate in the highest frequencies (i.e., the ultraviolets). This is true because higher frequencies carry higher energy; therefore, when you add energy to the blackbody in the form of heat, the frequencies given off as light radiation should increase in proportion to the energy put in. It all seemed perfectly clear and explainable under Newton's centuries-old equations of classical mechanics. The trouble started when blackbodies refused to act as predicted when they were taken into a laboratory and heated.

In practice, all bodies—no matter how efficiently they radiated—were found to have a "hot spectrum," with a peak somewhere in the infrared region. This was at the opposite end of the spectrum from that predicted. Nobody was ever able to find experimental evidence for waves predominating in the higher frequencies. As time wore on, this became known as "the scandal of the ultraviolet." In 1899, Max Planck solved the scandal and in so doing invented quantum mechanics. He pointed out that classical mechanics relied on the assumption that a perfect radiator would emit and absorb radiation of all frequencies smoothly, in continuous streams, and could do so in quantities of any size, however small. Planck replaced this with his own, entirely new assumption: Radiation could be given off or absorbed only in quite definite minimum amounts (which he called *quanta*). Nature acts less like a fountain dispensing reality in a smooth flow and more like a gum-ball machine which dispenses energy in tiny, discontinuous packets.

Blackbody Radiation—Newton's Puzzlement

Not only did Planck's explanation fit the experimental data, it led almost immediately to a global rethinking of the way physics modeled waves and particles. The core of quantum theory involves the counterintuitive notion that a wave can be a particle and a particle can be a wave. The very nature of physical reality, and the limits of what we can know about it, rest solidly on the wave-particle paradox.

CONVERSATIONAL TACTICS:

Nobody experiments with blackbody radiation anymore—it's a problem that was solved eighty years ago. Many people will think it is simply an obscure chapter from the history of physics. This may be so, but it is just that obscurity which will capture attention when talk turns to a subject that does happen to be *au courant*—the structure of scientific change. Prior to 1962, this was also a topic discussed only in academic seminars and other equally unexciting forums. However, in that year Thomas Kuhn published *The Structure of Scientific Revolutions,* and since then, the methods by which science progresses have been central to many arguments and debates outside the ivied halls.

Blackbody radiation is a perfect illustration of the kind of event that triggers a scientific revolution. Before Kuhn, the dominant view was that science progresses in a straight line, with everyone sharing the same assumptions. Kuhn changed all that by introducing the notions of *paradigm, normal science,* and *crisis.* Within scientific epochs, scientists share paradigms—systems of common assumptions about the meaning of problems they are considering. For example, before Copernicus showed the sun to be at the center of the solar system, the astronomical paradigm was the Ptolemaic view that the earth was the center of the universe. Before Planck, the physics paradigm was that waves and particles were distinct entities.

Within the paradigm, normal science takes place—the working out of unsolved problems. Crises (like the scandal of the ultraviolet) arise when normal science cannot solve those seemingly minor problems. When this happens, the old paradigm is eventually overthrown by the newer, better explanation (like those of Copernicus and Planck). Scientists then accept the new paradigm and begin a new round of normal science. Seen in this way, good science is more revolutionary than conservative.

The next time you hear talk of scientific change, paradigm shifts, or other socio-philosophical esoterica, don't fall back on the old saws (Ptolemy versus Copernicus, Columbus versus the flat-earthers). Show a little class and use the example of Planck and the blackbody problem.

RELATED TOPICS:

Quantum Theory, Randomness, Relativity, Uncertainty Principle.

FURTHER READING:

The Structure of Scientific Revolutions, T. S. Kuhn, University of Chicago, 1970.

Black-body Theory and the Quantum Discontinuity, T. S. Kuhn, Oxford University Press, 1978.

BLACK HOLES

DEFINITION:

Any region in space massive enough to prevent the escape of light waves.

WHAT IT REALLY MEANS:

The greatest enigma about black holes is not so much that they cause us to totally rethink what we mean by space and time, but that the more we find out about them the more confusing they seem. *Talking Tech* has organized the discussion of black holes in terms of "B.A.," "M.S.," and "Ph.D." levels of description. Keep in mind as you ascend through the levels of explanation that sometimes knowledge is confusing, and more knowledge is just more confusing.

Black holes are most easily understood if they are viewed as the outcome of a merger between theories of stellar life cycles and the theory of relativity. Stars are nothing more than huge balls of gas. During their lifetimes, they are kept in a sort of equilibrium by the outward pressure of the thermonuclear reactions in their core and the inward force of gravity. When a star's "fuel" (helium and hydrogen) becomes exhausted, it must contract under gravitational pressure, which needs no fuel except mass. If the star was massive enough in life—greater than twice the mass of our sun—then gravity will be so strong that the entire star will eventually be squeezed down to a radius of about two miles, producing an object so massive, with a gravitational field so intense, that not even light can escape its grip. As a result, the collapsed object is totally black to any outside observer. It has become invisible.

The master's degree level of explanation not only complicates the black hole issue, it confuses the nature of space-time in a serious way. According to Stephen Hawking, the acknowledged guru of black hole theory, the universe may have trillions of *mini* black holes which actually spew forth particles, erupting as *white* holes. The idea of "black" in black holes makes a very specific type of sense to physicists—it means that information (light) is forever withheld from the observer. Hawking's equations suggested a Zenlike turnabout on black hole theory, an idea that he often puts in the form of a riddle: "When is a black hole not black? When it explodes!"

The aspect of mini-holes that seems most paradoxical is that they not only suck in matter, radiation, and mass—they also spew some forth! This is described by the Hawking radiation process: Energy may be converted into matter (that's good old $E = mc^2$ read backward—like cramming an A-bomb blast into a piece of solid plutonium). Black holes have tremendous gravitational energy, therefore we should expect them to create particles of matter. Hawking has been able to develop a mathematical theory which predicts the level of particle production for specific sizes of black holes.

The larger the black hole, the cooler it is, and the particle production is almost nil. Mini black holes, however, are much hotter, and according to the theory some of them should be exploding right now, gushing both X rays and gamma rays and turning into white holes. Far from being just the final state of large stars, black holes seem to have a life cycle of their own. Are white holes the afterlife of black holes from another universe?

Mini black holes could have resulted as a consequence of the big bang that created our universe around fifteen billion years ago. The enormous energy of that singular burst could easily have squeezed small bits of space-time until they had a mass of a billion tons in an area the size of a *proton* (10^{-13}cm). Hawking thinks that these primordial holes may litter the universe, creating a Swiss-cheeselike effect. Decades ago, Einstein and Nathan Rosen wrote a paper about similar "wormholes" in space-time. J. Robert Oppenheimer was tackling the problem in the late 1930s, before he was called to the Manhattan Project. Some of the most dazzling minds in the history of science had skirted or toyed with the possibility of mini black holes before Hawking came along and turned things inside out.

Dealing with black holes at the Ph.D. level is as much an exercise in imagination as it is an exploration of physics. Once a sufficiently massive star starts to collapse, there is no known force in the universe strong enough to stop gravitational contraction. Even past the two-mile radius (the threshold for capturing light), the mass continues to collapse until it becomes an "object" of infinite density and zero volume, called a *singularity.* At this point, all the predictive machinery of modern physics threatens to break down. No one knows what goes on inside a singularity; thus far, they are considered cosmic free-fire zones. Time may run backward or sideways. Space may puncture or shred in unknowable ways.

The best bet so far is that space and time are badly warped inside a singularity, and that our everyday notions of cause and effect, past and present, here and there, would no longer do us much good. As Einstein pointed out, space is not flat; it is curved. Since space is also in some sense material, it can be curved even further by strong gravitational forces, which is exactly what a singularity is, if it is anything—a strong gravitational force. Imagine space in your mind's eye as a rubber sheet; the singularity is a heavy point pressing down on the sheet, distorting it, causing it to curve.

Einstein also showed that there is no such thing as absolute time; rather time is relative to a specific spatial framework. Time is one of the ways space happens; hence, as space is warped, so may be time. If it were possible for a human to exist at a singularity, you literally wouldn't know if you were coming or going, but it wouldn't matter, since you wouldn't know if you were here or there.

CONVERSATIONAL TACTICS:

Hawking has attempted to unite the laws of gravity (which usually only apply on the largest scale) with the laws of quantum physics (which usually apply only on the smallest scale) into a theory of quantum gravity which could explain what happens in black holes. This result would not make Einstein very happy. He argued against the probabilistic nature of quantum theory in the 1930s with his famous quote: "God does not play dice with the universe." Niels Bohr, Einstein's colleague and friend, one day countered with: "Albert, stop telling God what he can and cannot do." Because black holes are, by definition, invisible, Hawking recently coined his "principle of ignorance": "God not only plays dice, He sometimes throws them where they can't be seen."

Although there isn't much you can say with certainty about the nature of a black hole, there are a couple of reasonable answers available to obvious questions. The next time your child comes home from school or the movies with a burning desire to know something about black holes, you'll be prepared.

Q: What would happen if I got sucked into a black hole?

A: First, you would be stretched out like a noodle, because the part of you entering the hole first would have more force applied to it; then your body would break apart, and then your body's atoms would disintegrate until you were just a bunch of disconnected subatomic particles. You wouldn't feel anything, though, because it would happen faster than messages could travel from your sense organs to your brain.

Q: If black holes are invisible, how do we know there are such things?

A: Actually, we don't know for sure. Well-confirmed physical theories predict that black holes should exist, and scientists have recently begun to gather evidence of effects which seem as if they could only be caused by black holes. For example, a black hole should create a kind of gravitational whirlpool in space. This whirlpool is called an *accretion disk,* and the gravity of the black hole is so strong that it causes

Black Holes—God's Gamble

the disk to spin very quickly, heating up to such a high temperature that it begins to emit large quantities of X rays. Astronomers have now identified four X-ray sources in space which seem to be caused by black holes.

RELATED TOPICS:

Big Bang, Cosmology, Energy Alternatives, Radio Astronomy, Relativity Theory, Stellar Death.

FURTHER READING:

Black Holes and Warped Spacetime, William J. Kaufmann III, W. H. Freeman, 1979.

Black Holes, Walter Sullivan, Warner Books, 1979.

CHOLESTEROL

DEFINITION:

A white, tasteless, odorless, waxy substance produced by the body, but also found in some foods.

WHAT IT REALLY MEANS:

Whether it's macrobiotic diets, or vitamin B-15, or liquid protein supplements, the nutritional components of tech talk seem to have a shorter life span than the laboratory-produced, radioactive elements. Cholesterol, thanks partly to Madison Avenue, which has used low-cholesterol and its running mate, polyunsaturated fats, as generic terms signifying "good for you," is a major exception. For about thirty years, the dominant medical wisdom was that cholesterol was a major contributing factor to arteriosclerosis (hardening of the arteries) and other cardiovascular problems. About five years ago, new studies demonstrated that there were different types of cholesterol, and that some of them, known as high-density lipoproteins (HDLs), might even be beneficial. Finally, in 1980, the prestigious National Research Council published a report declaring that cutting out fats and cholesterol really doesn't make that much difference in preventing coronaries. The course of this episode in the management of technological information teaches us as much about the politics of science as it does about cholesterol.

Scientists have known about cholesterol since 1812, when it was isolated from gallstones. The name comes from the Greek *chole,* meaning gall or bile, and *sterol,* the chemical family of lipids to which it belongs. The first medical link appeared a century later, when a Russian scientist demonstrated that he could cause arteriosclerosis in rabbits by feeding them large quantities of cholesterol. Ensuing work failed to link cholesterol to heart disease in humans, but it did uncover many facts about the substance. The average person's body contains six to eight ounces of cholesterol, which is mainly produced by the liver. This chemical, which is really a lipid (a kind of fat), is needed for sheathing nerve fibers, for producing sex hormones, and as a component of the bile used to break down fats. In 1940, cholesterol seemed to be just another one of the body's necessary substances.

It wasn't until just after World War II that cholesterol became medically interesting. As a result of wartime shortages, certain European countries had low-fat diets imposed on them, and researchers later noted that the rate of heart disease had fallen in these countries. This led to the cholesterol hypothesis: There is a definite link between heart disease and levels of blood (serum) cholesterol, and these levels are highest in individuals whose diet is rich in animal fats. In this country, where heart disease rose from a cause of 8 percent of deaths in 1900 to 28.4 percent of deaths in 1940 to 37.8 percent of deaths in 1975, the hypothesis became more like an iron law. Limited-cholesterol diets proliferated, cholesterol-free products hit the market, and advertisers jumped on the health bandwagon. Science and technology became an integral part of a marketing strategy.

The trouble with the hypothesis, in the opinions of more careful researchers, was that it failed to explain all the facts. There were simply too many counterexamples: Why were women seemingly immune to the effects of cholesterol prior to menopause? How could one account for all those octogenarians who had rich diets, high cholesterol levels, and no signs of cardiovascular disease? Why were some families totally immune to heart problems? The answer turned out to be that there were two kinds of cholesterol, and what was important was the *ratio* of the two instead of the absolute amount of the substance. The HDLs are the beneficial kind, and they seem to float around the bloodstream, picking up excess cholesterol and returning it to the liver for excretion. The bad kind, known as low-density lipoproteins (LDLs), pick up cholesterol and deposit it in the cells for processing. However, if there is more cholesterol than needed for daily metabolism, some of the LDLs may deposit their fatty cargoes on the interior linings of coronary arteries. This buildup is what eventually may lead to a heart attack. The hypothesis had changed: The greater the HDL level, the lower the chance of heart problems; the greater the LDL level, the greater the chance of heart problems.

In 1980, the roof caved in on the cholesterol hypothesis advocates and triggered a bitter volley of recrimination which could only have happened in today's politically charged sci-tech area. The National Research Council reviewed more than a dozen cholesterol studies and found that attempts to switch to a low-fat, low-cholesterol diet had "no effect on overall mortality." They also cited studies showing that cholesterol-lowering drugs had no significant effect on coronary deaths. Finally, they pointed to disturbing evidence that lowering cholesterols may actually increase the risk of cancer.

No sooner had the report been made public than it was hit by a double-barreled blast. One dissenting group of scientists objected to the report on technical grounds; they claimed that there was voluminous evidence linking cholesterol to heart disease, and that it is just bad science to deny it. A second group, composed mainly of nutrition lobbyists, adopted a more political attack. They pointed out that members of the panel were also paid consultants to the egg and meat industries, exactly the folks who stood to profit from such findings.

These arguments left the lay public totally confused about cholesterol but did enlighten them regarding two principles of science in the age of tech talk. First, the institutional character of science itself has changed dramatically. Many scientists now involve themselves in policy issues with ethical, economic, social, and political implications. These involvements naturally lead to disagreements, which eventually become public. Science is no longer practiced solely behind laboratory doors. Second, in tackling policy issues, scientists have opened the way for a closer public scrutiny of their methods. The effect of individual bias on the results obtained from research is certainly an old political problem, but it is a relatively new one for science and scientists. The furor over cholesterol may just be symbolic of what is to come—acrimonious scientific debates which shed much heat but precious little light.

CONVERSATIONAL TACTICS:

Food is always a potent topic of conversation. Whether it's the price of hamburger, a new recipe, or the most fashionable new restaurant, food is almost as popular as sex. Just as popular as food is the lack of it—

more precisely, the latest diet fad. You don't have to be a pollster to know that many, many Americans are either on a diet or talking about one at any given moment. Nor is this knowledge lost on the publishing industry. During the last ten years hardly a week has passed without some miracle diet appearing on the best-seller list. To find out how much attention you've paid to these purveyors of svelte, review the quiz below. If you score six or better and you still need to drop a few pounds, you should consider the possibility that you're doing too much sedentary reading. Stop reading diet books and get some exercise!

DIET QUIZ

Match the diet with its most prominent feature.

1. Doctor Atkins'	A. Longevity diet
2. Dr. Belham's	B. No-aging diet
3. Dr. Cantor's	C. 1-2-3 Sports diet
4. Dr. Cooper's	D. Superenergy diet
5. Dr. Frank's	E. Save your life diet
6. Dr. Greff's	F. Natural fiber permanent weight-loss diet
7. Dr. Reuben's	G. Virility diet
8. Dr. Siegal's	H. Water diet
9. Dr. Stillman's	I. Complete Scarsdale medical diet
10. Dr. Tarnower's	J. Fabulous fructose diet

ANSWER KEY: 1-D, 2-G, 3-A, 4-J, 5-B, 6-C, 7-E, 8-F, 9-H, 10-I.

RELATED TOPICS:

DNA, Endorphins, Longevity—Immortality, Recombinant DNA.

FURTHER READING:

Toward Healthful Diets, National Research Council, 1980.
Fact-Book on Fats, Oils, and Cholesterol, Carlson Wade, Keats, 1973.

CHRONOBIOLOGY

DEFINITION:

The study of rhythmic processes which take place inside living organisms.

WHAT IT REALLY MEANS:

Science and common sense do not always share the same view of the world. How many of us actually believe that this page is mostly empty space, or that our noblest impulses may be controlled by our biology, or that time would slow down if we traveled very fast? Even though society may occasionally protest (witness the fate of Galileo or the deaf ear turned toward Alfred Wegener, an early proponent of continental drift), we expect science to take the lead in shaping our world view. It is precisely this expectation which makes it all the more satisfying when scientists finally get around to "proving" something most of us knew to be true on the basis of our everyday experience.

The existence of an individual internal clock is just such an item. Anyone who has ever flown the "red-eye special" and expected to put in a solid day's work, or who lived with a roommate who was doing deep knee bends at six A.M. when you couldn't even pry your eyes open until after noon, knows implicitly that one's body and the clock don't always agree on what time it is. Now scientists are not only studying this phenomenon, they believe they may have a partial explanation for it.

Humans have always been interested in the rhythms of nature. The first scientist-priests were able to attain their status by predicting the changing of the seasons, yet it was not until the eighteenth century that a scientist was able to demonstrate that animate objects might have their own internal rhythms. The initial object of study was the heliotrope plant, an organism known to open its leaves in the morning and close them at dusk. Even when all environmental cues like light and temperature were removed, the plant maintained its cyclic behavior—

the first evidence for the existence of biological clocks. During the next two centuries, many species of plant and animal were found to exhibit similar behavior, and by the mid-1950s the basic tenet of chronobiology could be set forth: internal clocks are a fundamental property of all life. Scientific work then shifted to explaining the mechanisms of the clocks.

Remember the early TV commercials for Accutron watches? They showed a little vibrating tuning fork and proclaimed that the timepiece was kept accurate to within fifteen seconds a month by the oscillations of the tiny tines. Recent scientific work has discovered that we have something very similar residing in the hypothalamus—cells called the suprachiasmatic nuclei (SCN). The SCN are two tiny structures which actually pulse as the rate of nerve firings increases and then decreases. Their pulsing establishes our fundamental time cycle, which is transmitted to the rest of the body via neural and hormonal signals.

As dramatic as this is, most chronobiologists believe that we have more than one biological clock, and work is proceeding on the development of a general model of human biological timekeeping. One theory is that the opening and closing of our cell membranes in response to ionic concentrations may establish the basic rhythm. The competing theory hypothesizes that our biological time is intimately linked to the rate at which we manufacture proteins. When our cells' protein-making apparatus is disturbed, it also disturbs our internal clocks. Whichever theory, or combination of theories, proves correct, scientists and doctors are already applying what we know about chronobiology to everyday life.

Everyone is familiar with the prescription label's instructions, "Take three or four times daily." If chronobiologists are correct, the label may soon read, "Take when your window for drug effectiveness is greatest." Experiments with laboratory animals indicate that a drug's effectiveness varies, sometimes drastically, depending upon when it is administered in relation to the animal's daily rhythm. Such results could lead to a science of chronotherapy—the use of biological time to insure most effective treatment.

Chronobiology is also being used in a new diagnostic science—autorhythmometry. The object is to have each individual monitor his own vital signs (temperature, heart rate, blood pressure) six times daily and to develop a chart of his own body's rhythm. Taking measurements at different times is crucial because we now know that the body doesn't randomly fluctuate around its norm (98.6 degrees Fahrenheit) but rotates around it in orderly cycles: blood pressure is lowest just after waking, pulse rate is lowest during deep sleep and quickens before waking, skin cells divide most frequently during sleep. The clinical belief is that any sudden variation might be a clue that something is wrong.

CONVERSATIONAL TACTICS:

Dieting is a way of life in America. If we're not doing it, we're talking about it. Whether you need to shed a few pounds or simply want to chew the fat, here is the world's simplest diet—the chronobiological diet. In an experiment at the University of Minnesota, volunteers were given a single, two-thousand-calorie meal each day. When it was eaten at dinnertime, all gained weight or remained the same. When it was eaten for breakfast, they all *lost* weight. The message is clear—you are what you eat, but more so if you eat at night.

RELATED TOPICS:

Behavior Modification, Consciousness, Endorphins, Photosynthesis.

FURTHER READING:

"The Times of Your Life," Dianne Hales, *Next*, July/August, 1980.
"The Clock Within," Philip Hilts, *Science 80*, December 1980.

CLONES

DEFINITION:

An organism or group of genetically identical cells descended from a single common ancestor and produced by a process of asexual reproduction.

WHAT IT REALLY MEANS:

There's no need to be alarmed, but you are probably surrounded by clones at this very moment. In fact, if you've ever grown a houseplant from a cutting, you've nurtured a clone. Cloning of plants in this way has been common practice for centuries, and the word "clone" comes from the Greek word for twig. Each plant grown from a cutting is an exact genetic copy of the original.

Cloning didn't become scientifically interesting outside the world of horticulture until the 1950s, when two American scientists devised a method of transplanting the nucleus of a cell (the DNA-containing component). Theoretically, every individual cell in any living organism contains all the information needed to grow a new organism. What is missing in ordinary cells is a way to unleash that potential, to make the central planning molecule release the orders to build a new carrot, or mouse, or human. A breakthrough in laboratory method provided a first, rough means to trigger that information.

The "nuclear transfer" technique enabled the scientist to transplant the genetic material (nucleus) of the organism to be cloned into an egg cell of a nucleus-depleted host. This microsurgical manipulation allowed the clone cell to grow and divide without mixing with any of the host's DNA. If the clone cell were able to mature, a genetically identical adult organism would result; however, in the 1950s, growing a mature clone was just science fiction.

Science fiction crossed the border into science fact in the early 1960s, when Cornell biologist Frederick Steward succeeded in cloning a carrot plant from a single carrot cell. What's more, he didn't have to use the nuclear transfer technique because plant cells are *totipotent,* meaning they are able to express their entire genetic endowment. This meant that Steward didn't need an egg cell; instead, the cloned cell could be placed directly in a culture nutrient and it would multiply, like a bacterium. At about the same time, British biologist John Gurdon used nuclear transfer to clone frogs from tadpole cells. In 1975, he was able to clone a frog from the cells of a mature frog. This was critically important because it proved that even cells which have undergone *differentiation*—which have changed in form to meet a specific function—still have all the genetic information needed for cloning.

Why would anybody want ten thousand genetically identical carrots? To feed those ten thousand genetically identical frogs? Sexual reproduction produces variation, but cloning produces genetic constancy, and therein lies its value. Through the cloning process, scientists can produce pure strains of laboratory animals to meet any experimental need. Using similar techniques, scientists could select a genetically superior plant and grow any number of exact copies. Cloning of livestock may lead to industrial mass production of meat. Instead of beef on the hoof, the carnivores of the twenty-first century may eat what started out as beef on the petri dish.

The fundamental question, of course, is the scary one: Can humans be cloned? Will you or I ever be able to clone ourselves and thus marry narcissism and biotechnology? Right now, the answer is somewhere between "yes" and "maybe." While scientists have only begun to clone mice (mammals are more difficult than frogs because the cloned cell must be placed in a womb), there is no reason, in principle, why cloning a human should be impossible.

Clones—Feeding Time at the Frog Farm

Within each cell of every human being are forty-six chromosomes that specify every detail necessary to grow an exact replica of that body. In sexual reproduction—the normal mammalian way of doing our evolutionary duty—two incomplete cells, with twenty-three chromosomes each, combine to make a new person. To reproduce a human by cloning would require only an ovum and one cell from any part of a living person, male or female. By destroying the nucleus of the ovum and replacing it with a full forty-six-chromosome nucleus removed from another cell, the technician might be able to trick the ovum into growing an embryo with the transplanted nucleus in place of its own; the embryo could then be implanted in a host mother and brought to term, exactly replicating the human who provided the nucleus.

No scientist has yet published or dared talk about the results of a successful, authenticated human cloning, although a writer named David Rorvik published a controversial book *(In His Image: The Cloning of a Man,* Lippincott, 1978), that he claims to be the account of a cloning accomplished for an anonymous aging millionaire by a secret team of scientists in an unidentified country. Whether or not Rorvik's unknown clone actually exists, many reputable researchers feel that human cloning technology is inevitable.

Whether humans *can* be cloned and whether they *ought* to be cloned are different questions, and the ethical implications of this potential technology promise to make the abortion controversy look tame by comparison. For example, if it is morally improper to grow an unconscious replica of yourself in order to ravage it for spare parts, would it be any more proper if the replica is grown without a brain? Is it ethically permissible to prolong your life by altering a few of your own cells to grow your own organ warehouse? Questions such as these are part of the thorny new field of *bioethics.*

Aside from ethical considerations, there may be medical risks to cloning yourself. Recessive genes may express nasty traits such as hemophilia in your clone, even though they don't show up in you. Cloning will not provide complete control of the embryo; the ectoplasm surrounding the nucleus may set the program for early embryonic development. Finally, there is the poetic problem Lewis Thomas pointed out: "Cloning is the most dismaying of prospects, mandating as it does the elimination of sex with only a metaphoric elimination of death as compensation."

If you or your co-conversationalists are still interested in self-cloning despite the risks and philosophical problems, you might want to consider the experiences of identical twins. Twins are naturally produced when a single fertilized egg splits into two cells, each of which develops into an embryo. They are, therefore, genetically identical, not with either one of their parents but with each other—twins are "self clones." Twinhood helps if you need an organ donor, but most psychological studies show that identical twins are no more and no less happy than you and I.

CONVERSATIONAL TACTICS:

Clone jokes are presently rather passé, and when a true human clone comes along, we might not be so willing to joke about it. Instead of a clone joke, *Talking Tech* offers a *practical* joke, an inexpensive prank to pull on your host or hostess after you have stayed in their home. When you return to your own home after your visit, send them a postcard with the following words, substituting your own and your host's names:

"Dear Host,

We at CLONES UNLIMITED are pleased that you enjoyed the recent visit of our latest guest clone. The clone reported no malfunctions during your brief encounter. We have begun the new host clone from toothbrush scrapings of the host's toothbrush, which were clandestinely gathered by our guest clone. The process takes approximately ninety days, at which time the host clone will be sent to visit for your approval. If you are completely satisfied, there is an easy credit plan for purchase, which may be arranged at that time with the host clone. Thank you for your continued interest in CLONES UNLIMITED."

RELATED TOPICS:

Bioethics, DNA, Origin of Life, Recombinant DNA, Reproductive Technology, Sociobiology.

FURTHER READING:

Who Should Play God? Ted Howard and Jeremy Rifkin, Dell, 1977.
The Medusa and the Snail, Lewis Thomas, Viking, 1979.

COGNITIVE DISSONANCE

DEFINITION:

A psychological theory which claims that a state of tension arises when a person holds two thoughts, beliefs, or attitudes (cognitions) that are logically inconsistent (dissonant). Since the tension is perceived as discomfort, the affected person is motivated to reduce the discomfort by behaving in a way likely to reduce the dissonance and restore cognitive balance.

WHAT IT REALLY MEANS:

Cognitive dissonance is really just a fancier, more specific name for what everyone used to call rationalization. It occurs when someone's perception of the merit of one thing (a car, a job, a suitor) is changed by the person's strong commitment to it (purchasing the car, accepting the job, marrying the suitor). The theory also relates to the notions of bias and conflict of interest. What is scientific about cognitive dissonance are the actual experiments conducted by psychologist Leon Festinger and others—experiments which produced real data about how people act when they hold inconsistent beliefs. Festinger's theory accounts for that observed behavior and predicts how most people would act in similar situations.

The first series of experiments dealt with decision-making and demonstrated that when people were forced to make a choice between two equally attractive alternatives, they would subsequently find ways to enhance the good points of their choice while playing down its bad points. Conversely, they would exaggerate the bad points of the rejected choice and minimize its good points. It seems that once a decision is made, the grass no longer looks greener on the other side.

Teenage girls and pop records may sound like unorthodox elements for a serious psychology experiment, but Jon Jecker of Stanford University provided experimental support for dissonance theory by asking high school girls to rate the attractiveness of each of a dozen records. Two records that she rated as only moderately attractive were then offered to each girl, and she was asked to choose one of them as a gift. After she made her choice, each subject was again asked to rate the twelve records. Dissonance reduction occurred as predicted—the previously unattractive records took on added value when they became gifts. By manipulating the experimental sequence, Jecker also demonstrated that dissonance reduction does not occur during the process of making a decision, but only after the decision is made.

Another set of experiments dealt with cases in which a person is forced to make statements that differ from privately held beliefs (what we commonly call lying). Festinger found that people generally act to reduce dissonance by changing their beliefs to match their public statements. Perhaps this is what the politicians have in mind when they say: "I didn't lie, the truth just changed."

The final set of experiments were conducted by Festinger, E. Aronson, and J. Carlsmith, and this investigation focused on the phenomenon of temptation. The experimenters showed that when you or I want something and discover that it is unattainable, two different situations can result. One, if the deterrent to getting the object is weak, the "sour grapes effect" takes over, and we tend to derogate the previously desired object. Or, two, if the threat or obstacle standing between us and the desired goal is a very strong one, we have less of a tendency to reduce dissonance and often find the goal even more attractive. This last bit of information should be useful to parents when they have to decide how to say no to their children.

CONVERSATIONAL TACTICS:

Cognitive dissonance is a crucial concept for your conversational repertoire. Next time your loudmouthed neighbor starts in with his loving descriptions of his new car or vacation or swimming pool, just wink and say: "That's your dissonance reduction speaking." If you get stuck on a bar stool next to some old leatherneck who insists that boot training was the best experience of his life, tell him he's suffering from an acute case of initiation justification—the limiting case of cognitive dissonance.

Why take Festinger's or *Talking Tech*'s word for the power of this phenomenon? Especially when you can make some mischief and prove it to yourself with a simple, homegrown, social psychology experiment: Bring home two possible gifts for your spouse and ask him/her to discuss the gifts' merits and demerits; then tell your mate that only one gift may be kept while the other must be returned. After the choice is made, engage your partner in another round of gift evaluation. If the theory is correct, his/her remarks should shift in favor of the chosen gift.

RELATED TOPICS:

Behavior Modification, Consciousness, Placebo.

FURTHER READING:

"Cognitive Dissonance," Leon Festinger, *Scientific American*, October, 1962.

Perspectives on Cognitive Dissonance, R. A. Wickland and J. W. Brehm, Halsted Press, 1976.

CONSCIOUSNESS

DEFINITION:

Investigation and manipulation of physical processes associated with awareness, cognition, associative memory, abstract thought, symbolic communication, and language.

WHAT IT REALLY MEANS:

"Studying the mind with the mind is like making a mirror by rubbing two stones together." This old Zen proverb succinctly states the paradox at the center of every scientific investigation of human consciousness. Philosophers have argued about this riddle for millennia, but early in this century, the uncertainty principle (see Uncertainty Principle) established the inseparable bond between observer and observed, and it was stated in terms of physics, the least metaphysical science. The oracles of quantum mechanics had something very strange to say about the study of the mind: *all* science, after Heisenberg, became the study of organized consciousness. Consciousness investigating itself has always been a notoriously circular affair, and modern science is no less immune to this trap.

Ask a neurophysiologist where consciousness resides, and she will talk of neocortical nerve tracts. Ask the same question of a psychopharmacologist, and he will reply, with equal assurance, that the mind is a broth of peptides and neurotransmitters. Behaviorists will ignore the question and expound on stimulus-response patterns. Other specialists look at consciousness and see electromagnetic fields or holographic information retrieval. And if you make the mistake of asking a group of philosophers, they will reverse the inquiry and ask if you are referring to the brain, the mind, the body, the model, any combination thereof, or none of the above.

The study and manipulation of consciousness, besides being conceptually tangled, is further complicated by social, ethical, and political issues. The use of technology to control human behavior, especially by government agencies, is a very concrete and continuing consequence of basic research. The religious and cultural implications of altered states of consciousness, the issue of parapsychology, the bioethical complexities of human experimentation, and the use of mind-control techniques by various cults spice the already boiling consciousness brew.

Despite the absence of a unifying paradigm, it is possible for the average person to grasp the basic principles of consciousness research. The easiest way to approach consciousness science is through the physical structures of the brain, by examining what is known about the origin and functions of the special tissues which manifest that elusive quality known as mind. Fortunately, there now exist two powerful models—the *triune brain* and *split brain* models—which simplify the mass of data concerning the human brain.

The triune brain theorists see our brains as the product of billions of years of evolutionary development. Consciousness, in this scenario, is the latest act in a very long-running drama. By comparing fossil evidence of the biological history of the human brain with what is now known about its physiology, these scientists have come to the startling conclusion that *Homo sapiens* possesses not one brain, but *three.*

The oldest and deepest brain structure—known as the reptilian brain or *R-complex*—is a living legacy of the dinosaur era, and it still governs our most instinctive, unthinking, ritualistic behaviors. The next oldest structure, which surrounds and interconnects with the reptilian core, is the *limbic system,* the patrimony of our mammalian ancestors. There is evidence that this second system is the seat of emotion and central command post for rage, love, nurturing, and fear. The newest addition to the evolving brain, the third layer, is the *neocortex.* If consciousness is to be localized in any one area of the brain, all the evidence indicates that it will be in this relatively thin coating of gray matter.

The triune brain theory was originally elaborated by Dr. Paul MacLean, director of the Laboratory of Brain Evolution at the National Institute of Mental Health. *Talking Tech* is taking the liberty of condensing the scientific evidence into an ultrafast time-lapse "thought movie": Imagine that you are watching the history of the brain flashing by at a couple of trillion times normal speed, starting from the time when some prehistoric fishes began to grow a bulge at the top of their spinal cords. It took only a few hundred million years for this bump of nerve tissue to evolve into the combination of spinal cord, hindbrain, and midbrain which came into its own during the reign of the dinosaurs.

"Eat this; flee that; mate now" pretty well sums up the dinosaurian way of "thinking," a pattern of behavior that prevailed longer than we humans would care to admit. It was a nice basic brain, capable of many tricks modern science still cannot duplicate. It kept the heart beating, the lungs breathing, the mouth chewing, and the species reproducing. The reptile brain ruled the world for hundreds of millions of years.

On the planet of the dinosaurs, 150 million years ago, the early mammals were barely noticeable, but they were already evolving new brain tissue to perform new functions. Perhaps it was because mammals had to nurture their young, or because fear, rage, and other emotional behavior were survival necessities for those not weighing thirty tons and possessing armor plating—whatever the reason, the mammals began to grow a new brain layer. The limbic system (limbic means "to grow a border around") gave the mammals an edge on the reptiles. Like an old house which adds rooms as the family expands, the evolving brain retained all of the older "reptilian" brain structures while developing newer, more versatile structures around them.

The limbic system is a kind of switching station between thought, emotion, and action. It has outposts in both higher and lower brain structures, and it also controls the endocrine glands, which carry orders from higher to lower centers by releasing chemical messengers known as *hormones.* With the advent of hormones, living creatures began to experience sensations which approximated human emotion. The limbic system is still a very important part of the human brain, and when it malfunctions, all kinds of bizarre behavior can erupt.

When life evolved beyond the early mammals, only two special branches of the evolutionary tree grew the final, crucial layer of brain matter. The cetaceans (whales and dolphins) and the higher primates (apes and astronauts) grew a neocortex to take the helm of the reptilian-mammalian brains. Neocortex literally means "new rind," and that is exactly what it is—a thin, delicate, relatively recent rind of nerve cells. The highest center, the seat of reason, grew like the skin of an orange, surrounding the innovative organ that started, billions of years ago, as a bulge atop a primitive spinal column.

The neocortex has the power to override older command systems, but not in all situations and not at all times. Although it plays an important role in moderating our ancient carnivorous urges, the neocortex also produces symbolic language, abstract thought, and those other qualities we recognize as unique aspects of human consciousness. The mapping of neocortical correlates of consciousness is currently being attempted through the use of drugs, through biofeedback and electromagnetic processes, and through study of surgical procedures necessitated by brain ailments.

One of the most helpful current models originated from surgical-psychological studies and concerns the bilateral properties of the neocortex. Not only do we carry around three brains in our skulls, but it appears that the two *halves* of our newest brain are further subdivided, just as our consciousness appears to be divided into two complementary but often separate modes of thought.

What is consciousness? One useful way to rephrase the question is: "In what ways do our brains *know?*" Psychologists of the split brain or *bilateral* school call our attention to two modes of knowledge. One is the rational, linear, analytic way of thinking which enables us to read, write, and perform arithmetic; the nonrational, holistic, intuitive way of thought is the other mode, the part of your mind you use to recognize one face out of thousands or to compose a melody which never existed before.

The first evidence that these ways of thinking might be related to different sides of the brain came from research into the effects of human brain lesions. Strokes or injuries to the left temporal or parietal lobes of the neocortex often impair the ability to read, write, speak, and perform mathematical operations. Similar lesions in the right hemisphere tend to impair pattern recognition, musical ability, three-dimensional vision. In a careful series of observations of patients with specific brain injuries, Roger Sperry of Cal Tech and M. S. Gazzaniga of the State University of New York at Stony Brook developed the split brain theory.

By performing perceptual experiments with those patients whose cerebral hemispheres were surgically separated for medical purposes, Sperry and Gazzaniga presented a tentative map of bilateral functions. The dichotomy is by no means universal, but in general, the left hemisphere processes information sequentially, while the right half does it simultaneously from multiple inputs. The left hemisphere is more heavily involved with language; the right hemisphere is more involved in spatial orientation. Many tasks are strongly specific to one lobe or the other, and many more require careful coordination between the two ways of knowing. Drawing an accurate representation of an object, for example, calls on the right hemisphere for contour, proportion, and background and the left hemisphere for details and internal elements.

Now that you have looked at the brain from bottom to top, from the beginning of evolution to the present, and from the left to the right sides, it's time to remember that these are all metaphors for a phenomenon which cannot be totally analyzed and labeled under neat categories. The hemispheres can take over each other's functions; biofeedback has demonstrated how the conscious brain can learn to control deep-brain functions. All three brain layers and both halves cooperate to keep us alive and conscious. Consciousness itself is not broken into different levels or modes. Rather, consciousness seems to be a continuum of states, each of which requires a complex orchestration of chemical, electrical, and physiological events.

Undoubtedly, we will all have a clearer idea of what dreaming is, and how creativity happens, and why psychedelic drugs seem to mimic madness or religious ecstasy when scientists are able to define at least part of these activities in terms of currents and chemicals. What will it mean if and when the measurable essence of mind is held spellbound by peptide potions and enclosed in webs of bioelectric fields? Ultimately, the effects of consciousness research are impossible to predict, simply because major breakthroughs are likely to change the very way we think.

CONVERSATIONAL TACTICS:

Altered states of consciousness have been primary conversational topics ever since the first prehistoric farmers drank fermented grain and decided to have a ritual. Today, the range of consciousness-altering techniques is still centered on fermented grain beverages but has expanded into every area of science and technology. Mind-altering drugs, flotation tanks, psychoanalysis, biofeedback, meditation, yoga, running, sweating, and dozens of other ways and means to change the state of your mind are as close as the nearest bookstore, ashram, or pharmacy. Nevertheless, none of these brain-benders approaches the universality of mankind's most persistent and least understood altered state—dreaming.

You may not believe in drinking, you might not be able to afford psychoanalysis, and it's possible that you've never heard of biofeedback or fasting. One thing can be said with certainty, though: You dream every night, whether you know it or not when you awaken. Psychophysiologists, using brain-wave and other objective measures, have determined that every normal human being spends an average of ninety minutes each night in a dream state. The problem, in this culture, is that nobody is taught how to *remember* their dreams and what to do with them. Not one person in a thousand has learned how to dream correctly. Every consciousness consultant, from psychologist to cultist, now seems to have a new answer to this educational deficiency.

There are many competing schools of dream work, from Gestalt to occult, but perhaps the most exciting is the area of *lucid dreaming*—dreams in which you realize that you are dreaming and consciously manipulate the elements of the dream without waking up. Think of the possibilities: Faust sold his soul for Helen of Troy; you could have her company, or that of anyone you wished, for the price of a dream. Want to fly to Bali without a plane? Ever feel like trading jokes with W. C. Fields, talking finance with J. P. Morgan, or simply eating a superb dinner at Maxim's? If you learn the art and science of lucid dreaming, all these possibilities and more are open to you every night.

Lucid dreams sound like a dream come true, and they can be quite handy conversationally. You are free to use your imagination in describing the potential of lucid dreams, and if your conversational partners voice skepticism about the scientific basis, you can refer them to a scientific study which not only confirms the existence of this unique state of consciousness but proposes a five-step method by which the skill can be learned! In the January 1981 issue of *Psychology Today*, Stephen P. LaBerge reported on research conducted at Stanford. LaBerge discovered that it is possible to enhance one's capacity for lucid dreaming by doing mental exercises while awake; he personally reported as many as twenty-six lucid dreams a month using his procedure.

If you want to seize control of your subconscious scenarios, try practicing the five steps developed by Dr. LaBerge:

"1. During the morning, I waken spontaneously from a dream.

2. After memorizing the dream, I engage in ten to fifteen minutes of reading or any other activity demanding full wakefulness.

3. Then, while lying in bed and returning to sleep, I say to myself, 'Next time I'm dreaming, I want to remember I'm dreaming.'

4. I visualize my body lying asleep in bed, with rapid eye movements indicating that I am dreaming. At the same time, I see myself as being in the dream just rehearsed (or in any other, in case none was recalled upon awakening) and realizing that I am in fact dreaming.

5. I repeat steps three and four until I feel my intention is clearly fixed."

RELATED TOPICS:

Artificial Intelligence, Behavior Modification, Bioethics, Human Origins, Intelligence Enhancement, Origin of Life, Uncertainty Principle.

FURTHER READING:

The Dragons of Eden, Carl Sagan, Ballantine, 1977.
The Psychology of Consciousness, Robert Ornstein, Harcourt Brace Jovanovich, 1977.
Toward a Science of Consciousness, Kenneth R. Pelletier, Delta, 1978.
"Lucid Dreaming—Directing the Action As It Happens," Stephen P. LaBerge, *Psychology Today,* January, 1981.

CORIOLIS EFFECT

DEFINITION:

An apparent force on particles or objects in the atmosphere caused by the rotation of the earth under them. The effect results in the particles' motion being deflected toward the right in the Northern Hemisphere and toward the left in the Southern Hemisphere.

WHAT IT REALLY MEANS:

Even as children, we all have an intuitive understanding of the physics of rotating bodies. Remember the merry-go-round you sat on and begged your father to push? You learned quickly that if you sat at the edge, you'd go very, very fast, but if you sat at the center, it was as if you weren't even moving. Without knowing it, you had discovered a well-known physical principle: An object is carried with different velocities when it rests on different parts of a rotating body.

This principle can be visualized on the biggest rotating body we come into contact with—the earth. Standing at the equator, you will rotate along with the earth, moving 24,000 miles in one day at a velocity of 1000 miles per hour. As you move toward the poles (toward the middle of the merry-go-round), the time still remains the same—one day per revolution—but the circle of rotation decreases, so your velocity must also slow down. At the poles (the center of the merry-go-round), your velocity will be zero. When this principle of rotation is applied to moving bodies (missiles, winds) headed north or south, the Coriolis effect comes into play.

Imagine an intercontinental ballistic missile fired from the equator toward the North Pole. Because of its inertia, it will tend to keep the original velocity with which it was moving east at the time of firing (1000 mph). However, by the time it arrived over the U.S.A., it would be in a region where objects were being carried eastward by the earth at a much slower speed. As a result, the missile will drift eastward, to the right, as it speeds toward the pole. Of course, had the ICBM been fired toward the South Pole, it would still drift to the east, but this would mean a leftward drift. The Coriolis effect is one of the factors airplane pilots and navigators must reckon into their calculations.

The effect is also important in understanding natural phenomena. Although the complete life cycle of hurricanes is still unknown, the Coriolis effect is believed to play a crucial role. It bends the winds converging upon the center of a developing low-pressure system into a vast, counterclockwise spiral (in the Northern Hemisphere). The spiral then tightens until a delicate balance is reached—the winds can no longer go toward the center and are forced upward in a large, vertical corkscrew. The rise is marked by a cylindrical wall of clouds that can tower over fifty thousand feet. If the central air in the cylinder leaves via "the chimney," the low intensifies and the southeast coast boards up for a hurricane.

Anyone who remembers the shower scene from *Psycho* is familiar with the Coriolis effect's most mundane manifestation—the spiraling of water down the drain. Although it is a matter of some scientific dispute, most researchers now feel that it is the Coriolis effect which causes water in the Northern Hemisphere to form a counterclockwise vortex, while water in the Southern Hemisphere gurgles down the drain in a clockwise spiral. The Coriolis effect may not be revolutionary science or even very important to our daily lives, but it is one of the intriguing side effects of life on a giant, spinning sphere.

CONVERSATIONAL TACTICS:

Nowhere in the universe is there a greater apparent need for small talk than at a cocktail party. The next time that deadly silence creeps over the room and you feel obliged to speak up, tuck your tongue firmly in your cheek and give a scholarly explanation of the traffic flow at the party. Assuming you're north of the equator and the flow is counterclockwise, simply opine that it's the damndest thing, but cocktail party traffic seems to obey the Coriolis effect.

Should the traffic be moving in a clockwise direction, forget Coriolis and extol the virtues of that astute student of human nature, C. Northcote Parkinson. He claimed that party flow always moves left to right and gave a biological explanation: Since the heart is on the left side, defensive shields were held in the left hand and offensive weapons in the right. Because the weapon was usually a sword, this meant that the scabbard would be worn on the left and horses would have to be mounted from the left. Finally, since you didn't want to mount your steed while standing in the roadway, you moved the horse to the left of the road. From all this, Parkinson concluded: "Free of arbitrary traffic rules, the normal human being swings to the left."

If you are partying south of the equator and the flow is clockwise, either Coriolis or Parkinson will do as an explanation. If the flow is counterclockwise, you might suggest that C. Northcote wasn't *always* correct, or you might consider dropping the matter altogether.

RELATED TOPICS:

Mathematical Modeling, Sunspots, Weather Modification.

FURTHER READING:

The Ambidextrous Universe, Martin Gardner, Scribner's, 1979.
Parkinson's Law, C. Northcote Parkinson, Ballantine, 1957.

COSMOLOGY

DEFINITION:

The astrophysical study and theoretical modeling of the origin, structure, and constituent dynamics of the universe. The current standard cosmological model posits a homogeneous, isotropic, expanding universe of finite age.

WHAT IT REALLY MEANS:

Cosmology is at the same time the grandest and the least earthly of the sciences. Armed only with their equations, a few smudges of light on photographic plates, and the flimsy tracks of elementary particles colliding in bubble chambers, cosmologists attempt to construct models of the biggest thing there is—the universe. Yet such far-flung hypothesizing need not be totally inapplicable to life here on the planet of the humans. Don't forget that J. Robert Oppenheimer and his Manhattan Project colleagues were obscure professors engaged in speculations about esoteric stellar processes shortly before they gathered on a New Mexico mesa to give every earthling a lesson in nuclear physics.

The importance of cosmology transcends even the scariest technological potential, because the portraits that scientists paint of the structure of the universe are potent images for the whole human race. By comparing the history of science with the parallel history of human affairs, one cannot help but notice that the models we create regarding the size and age of the universe always alter the way we think about ourselves, our purpose, and our destiny. The Copernican revolution four hundred years ago produced such heavy social detonations that we still hear the echoes—empirical science, the industrial revolution, the rise of technology. We're all part of the secular, earth-centered, pragmatic, Copernican "new wave."

The Copernican wave crested a few decades ago. A new revolution is under way, and sooner or later its significance will ripple out to change the face of the world. It all started in 1923, when Edwin Hubble and

Milton Humason noticed that the spectral lines on their slides of distant celestial objects seemed to be shifted toward the red end of the spectrum—the famed red shift. This discovery led to Sir Arthur Eddington's hypothesis that the entire universe is undergoing a violent expansion in which every galaxy is rushing away from every other galaxy at a rate directly proportional to the distance between them.

When they explain the expanding universe model, cosmologists traditionally use a visual simile: Imagine the universe as the surface of a balloon. The galaxies can be represented by dots painted on the balloon's surface. As the balloon is inflated, the distance between each dot increases at a rate proportional to the distance between them. Continuing the simile, the "big bang" is the explosion that originally produced the energy to expand the balloon.

Since Hubble and Humason's discoveries, the big bang theory has become so widely accepted that it is known as the standard model. Its onetime competitor, the steady state theory, which posits a continual creation with no beginning and no end, has been virtually vanquished. There is, however, an interesting wrinkle to the big bang idea, known as the oscillating universe hypothesis: If there exists a critical amount of matter in the universe, the force of gravity should eventually slow the expansion (if not, the universe could expand forever). At the point of maximal expansion, a safe several billion years in the future, everything should begin to contract, perhaps until all the galaxies are crammed into a singularity again, at which time a new big bang could trigger another expansion. At present, cosmologists are undecided as to whether there is enough matter in the universe to cause this oscillation. Present observational techniques indicate that there is *not* enough matter, but many scientists feel that we simply haven't found it yet.

The case of the missing matter promises to be one of the hot cosmological topics in years to come. Some astrophysicists think that the unobserved matter could be undetectable because it is captive in black holes. Others have proposed that neutrinos, which saturate the universe, might possess a tiny amount of mass—in contradiction to current particle theory—which could also account for the missing mass.

CONVERSATIONAL TACTICS:

If you are engaged in blowing up a balloon with dots painted on it, you have an excellent opportunity to point out a startling parallel between ancient mythology and modern cosmology: The Hindu concept of the Breath of Brahma is astonishingly similar to the oscillating universe model. Everything begins and ends, in the mythical Indian cosmology, with a sudden expansion of energy, which goes through three-hundred-million-year cycles called kalpas, until Shiva dances to destroy the cosmos and starts it all over again. For a millennia-old theory that was developed well before the age of radio telescopes, the Hindu cosmology is amazingly accurate.

Because we insist on referring to the cosmologies of past civilizations as nothing more than myths, we therefore relegate their cosmologists to the role of mythmakers. Are we certain that *our* cosmologists are any better? How can we be sure that the big bang theory won't look as silly in two thousand years as Plato's cosmological dialogue *Timaeus* looks today?

If you find that you can't resist arguing the affirmative against an obdurate cosmological skeptic, you do have one potent weapon in your conversational arsenal: *the anthropic cosmological principle.* The basic idea is that the physical conditions of the universe are limited by the fact that they led to human life on earth, which is now observing those conditions! The hypothesis seems a bit weird, but it is a legitimate cosmological theory, entertained by some of our most respected scientists.

As a concrete example of how such a metaphysical notion fits the physical evidence, consider the question of the size of the universe. In an expanding cosmos, size depends upon age: "How big is the explosion?" is the same question as "How old is the explosion?" According to the anthropic principle, the age (and hence the size) can be estimated by the knowledge that conscious beings (us) are here to speculate about it. If the space-time radius were smaller than the present estimate of fifteen billion light years, the universe could not have existed long enough for the heavy elements essential to the evolution of life to be synthesized in the cores of exploding stars. If the universe were much bigger and older than the presently observed radius, the stars necessary to establish the conditions of life would have burned out long ago. In other words, our very existence as conscious beings is a physical constraint on the types of universes we can observe. As observers, we are more or less forced to find a universe about as big and old as the one we see around us.

Point out to your skeptical friend that studying the universe is a self-referring behavior which is obviously observer-dependent. Add that the anthropic cosmological principle insures that current theories are at least in the right ball park. If none of this is convincing, invite your friend out for a look at the stars. They have been known to induce a sense of wonder in skeptics since time immemorial.

RELATED TOPICS:

Big Bang, Galaxy, Quasar, Radio Astronomy, Red Shift, Stellar Death.

FURTHER READING:

"Structure of the Early Universe," John D. Barrow and Joseph Silk, *Scientific American,* 1980.

Cosmos, Earth, and Man, Preston Cloud, Yale University Press, 1978.

DIGITAL COMPUTER

DEFINITION:

Any mechanism capable of solving mathematical equations or manipulating data through the discrete representation of variables.

WHAT IT REALLY MEANS:

For the thirty years immediately following World War II, television was the dominant technology. It even gave rise to an epochal sobriquet—"the TV generation"—which became a code phrase for everything that was wrong with society. Declining school performance, increased violent crime, disintegration of the family, and poor voter turnout were but a few of television's putative evil effects. Although few realized it at the time, the first year of network broadcasting, 1946, was also year one of the computer revolution—a technology which would grow so fast and far that all aspects of our lives, including television, would come to rely on it.

However, unlike television, the digital computer is praised as much as it is faulted. This technology is the perfect example of the trade-offs necessitated by the social adoption of scientific discoveries. Each potential benefit of the "computer society" is balanced by a perceived risk: increased leisure time versus loss of jobs to automation; improved productivity versus mind-boggling computer errors; increased ability to communicate versus depersonalization of society. That computers will continue to become more powerful is undeniable; the important question is how that new power will affect our lives.

The first digital computers were illegitimate children of World War II. Norbert Wiener is generally credited with the invention of the mathematical subspecialty known as *cybernetics,* which describes the actions of complex systems from electrical circuits to human brains. Cybernetics originally grew out of the war-research efforts of Wiener and Warren Weaver, who were thinking about ways to make antiaircraft guns predict the future. Back when planes were slower, you just aimed your ack-ack the way you aimed a rifle, as if you were shooting a faster-than-usual duck. In the age of the Messerschmitt and the Zero, you had to deal with high-speed evasive maneuvers. You needed a gun that could think ahead.

Ballistics, naturally, was a hot topic during the war. Vannevar Bush supervised the construction of a gigantic, impossibly unwieldy, electromechanical computer known as the differential analyzer in order to solve complicated ballistic equations. (They say it sounded like "a million knitting needles," because of the enormous number of relays clicking on or off at any time.) With the Nazis giving London deadly lessons in missile guidance systems, the novel idea of bullets which could aim themselves became a vital matter.

Several separate teams raced to build the first electronic computer. Aiken at Harvard was working on the MARK series. Goldstine, Mauchly, and Eckert labored at the University of Pennsylvania. John von Neumann, who is credited with the key concept of flexible programming, commuted between the Institute for Advanced Studies at Princeton and that ultrasecret summer camp for scientists down at Los Alamos.

The first electronic computing machine, known as ENIAC (Electronic Numerical Integrator and Calculator), was built at the University of Pennsylvania for the army. Its sole purpose, at first, was the computation of trajectories for new weapons. ENIAC was as big as a small house, weighed about thirty tons, and contained eighteen thousand vacuum tubes, the big bulky kind you find in old radios. It needed so much electricity that the story (perhaps apocryphal) went that when ENIAC was plugged in, all the lights in Philadelphia dimmed. But it was a marvelous computer; ENIAC could do in thirty seconds what it took an ordinary calculator twenty hours to accomplish.

The reason ENIAC was a computer and not a mere calculator was that it met a number of the criteria put forth in 1940 by Norbert Wiener, a set of rules which became known as the "computer blueprint":

1. The central adding and multiplying apparatus of the computing machine should be numerical.

2. The key mechanisms, which are switching devices, should be electrical rather than mechanical.

3. The numerical base of the machine should be binary rather than decimal.

4. The machine should contain all the logical circuits needed to process the inputted data into the desired output form.

5. The machine should be able to store data.

Understanding these five criteria will enable you to understand, in large measure, how a computer works. Wiener's first criterion requires a digital, rather than analog, machine; that is, it requires a machine that *counts* rather than *measures*. Instead of dealing with continuous quantities, the way a speedometer does, digital computers use discrete symbols to represent alphanumeric quantities. While a speedometer can read somewhere between 54 and 55, a digital speedometer must read 54 or 55. The second requirement simply separates high-speed electronic computers from their slower, mechanical cog-and-wheel cousins, such as old-fashioned cash registers.

The third requirement, that computer arithmetic be binary, is one source of the general public's confusion and anxiety about how computers work. In a decimal system based on 10, each digit of a number, read from right to left, is understood to be multiplied by a progressively higher power of 10. Yet because computers are essentially systems of on-off switches, an arithmetic based on 2 makes more sense.

In such a binary system each digit of a number, read right to left, is multiplied by a progressively higher power of 2. The most important fact to remember is that any number in either system can be translated into the other, using a mechanical procedure. Humans can continue to use the decimal system, and the computer will switch the input—say, 30—into binary notation—11110.

The fourth criterion is the "hands off" clause. It says that computers should have the ability to process information independently of the operator—they must be able to carry out lists of instructions without outside intervention. Modern computers are composed of three main parts: an input-output device (I/O), a central processing unit (CPU), and a memory. The I/O can be a typewriter, a light pen and cathode screen, a remote sensor, or any other device which can translate human or mechanical input into machine-readable terms. The task of the CPU, known as a *microprocessor* when it is integrated on a single silicon chip, is to receive the inputted data, store it, perform arithmetical and logical operations in accordance with previously stored instructions, and deliver the processed information through an output device. The CPU is the "guts" of the computer; it's where the computing takes place. Wiener's final criterion, the ability to store data, is embodied in the computer's memory, a device for retaining binary-coded information until it is needed by the CPU.

If computers are nothing more than electrical boxes which manipulate digitized data according to prescribed programs, how can one account for their ubiquity and profound impact on society? There are actually two answers to that question: First, in the words of John von Neumann, a founder of automata theory, the computer is "the all-purpose machine." Second, technological advances have made the machine so small and inexpensive that complex, programmable microcomputers can now fit into the palm of your hand, and special-application computers can be found in your car, your television, your watch, and virtually any other mechanism where there is a need to control a process.

Computers are also all-purpose machines because they are incredibly fast—a standard large computer can make a computation in one billionth of a second—because they can store very large amounts of data and retrieve selected pieces almost instantaneously, and because computer specialists are always finding new ways to enlarge the scope of possible computer tasks. Speed and capacity are the results of improvements in hardware design and miniaturization. Programming advances are the result of progress in *software* design. In the language of the computer specialist, software is anything you can send over the telephone lines; hardware is everything else.

Ultimately, computers are so pervasive because they are the greatest means ever devised for handling information, and virtually all processes, from automating an industrial plant to teaching a child to read, may be understood in terms of a flow of information. As society grows more complex, and as data and information become more important to collective and individual survival, the role of the computer will continue to expand.

CONVERSATIONAL TACTICS:

Everyone has his favorite story about the computer run amok, making life miserable for us mere mortals. For example, there's the story about the vacationers who return to their unoccupied home to find a three-thousand-dollar utility bill. So many of these stories have made the rounds that *Talking Tech* suggests you turn the tables the next time the topic comes up and relate these "man bites computer" tales:

1. Computer crime has become a serious problem. According to experts, this type of larceny is far from petty—at least one hundred million dollars a year are lost in crimes of this type, and as criminals go, the computer criminal is extremely resourceful.

One of the earliest and still ongoing forms of computer crime is the use of a "blue box" to make long-distance phone calls. In the 1960s, a techno-outlaw known as "Captain Crunch" and other members of the "phone phreak" cult discovered that computers held access to long-distance phone lines and that the proper entry codes were composed of audible beeps and whistles. By building boxes which could emit the proper sounds over the telephone, the phreaks duped the communications computers, and students all over America were calling home, or Amsterdam, or Bombay, gratis.

2. Take a close look at your check deposit slips. Notice all those computerized numbers and codes at the bottom. That's all the computer is interested in, not the fact that your home address is printed on the form. One industrious computer criminal went around to the branches of his bank and left a number of his own coded deposit slips on the banks' counters. He reasoned that people wouldn't pay attention and just use the first slip they got their hands on and that he would get rich. He was right. He built up a mini-fortune before he was caught.

3. One East Coast programmer for a bank came up with a simple scheme that seemed undetectable. He created a special account in his own name and wrote an instruction in the bank computer's programming that all fractional cents created by rounding off all other accounts be transferred to his special account. In no time at all, the cents became thousands of dollars. But he too was caught.

Not all computer crime is so brainy. There is also the case of the disgruntled computer-error victim who walked in and plugged the machine with a revolver.

RELATED TOPICS:

Analog Computer, Appropriate Technology, Artificial Intelligence, Fiber Optics, Image Enhancement, Information, Miniaturization.

FURTHER READING:

Computer Power and Human Reason, Joseph Weizenbaum, W. H. Freeman, 1976.

The Computer Survival Handbook, Susan Wooldridge and Keith London, Gambit, 1973.

Crime by Computer, Donn B. Parker, Scribner's, 1976.

The Home Computer Revolution, Ted Nelson, published by the author, distributed by The Distributors, South Bend, Indiana, 1977.

DNA

DEFINITION:

Deoxyribonucleic acid, the molecule responsible for transmitting genetic information and regulating cellular life processes.

WHAT IT REALLY MEANS:

There is something about life that transcends our individual lives, an invisible force that insures renewal amidst corruption. We feel ourselves age and slide toward mortality, yet we recognize our grandparents' faces in our children's features. *Something* is being passed through time from body to body, like a precious baton in some grand and ancient relay race. For millennia, the great religions have been assuring us that the life spirit is an immortal language, an eternally self-renewing pattern of meaning, and the prophets have promised us that death is but an illusion, an artifact of our corporeal perspective. Thirty years ago, scientists started saying much the same thing, except the newest spokesmen for the doctrine of eternal life did not base their arguments on faith or revelation but on the shape of a molecule.

DNA, the famous "double helix" of James Watson and Francis Crick, is the world's best-known molecule. DNA's fame is well-earned, for more than one world-shaking reason. First, the importance of DNA's role in all living organisms cannot be overestimated—the infinitesimal collection of atoms is the immortal message of life itself. Second, and hardly less important, is the place the decoding of DNA holds in the history of science. In the long term, the deciphering of this molecule's message will be remembered by the human race as much more significant than even the splitting of the atom or the moon landings.

Most scientific questions are transitory phenomena—here one decade and gone the next. Only the most important and difficult problems linger for centuries, awaiting the proper solution. DNA is the final answer to a question first posed in the dim recesses of our prehistoric past: What is the mechanism of human renewal and the renewal of all life? Or, as we might say today: How does heredity work? To be sure, DNA was not the first answer proposed. The history of biology is littered with now discredited attempts: Aristotle's continuous action of psychic powers; a motley collection of theories embodying the notion of "vital forces"; *preformation*, which held that a tiny human, known as a *homunculus*, already existed in the head of the sperm cell; and *epigenesis*, the notion that the embryo gradually emerges from a formless mass. While each of these theories had its individual flaws, they all failed because they could not account for the different mechanisms needed to explain the renewal of life and transmission of hereditary traits from generation to generation.

So much has been made of DNA's shape that people tend to lose sight of its other remarkable properties. In particular, DNA is capable of performing the three tasks essential for the processes of heredity. First, it has a coding property; its subunits can be arranged in an almost endless variety of highly specific and unique sequences. Second, each DNA molecule can direct the formation of duplicates of itself; this is necessary not only for the passage of traits, but also for the building of the organism from a single cell into the trillions of cells which comprise higher organisms. Finally, DNA is capable of transmitting its coded sequence to a different nucleic acid (RNA), which can, in turn, direct the synthesis of specific proteins. In this way, DNA contains the blueprint for the corporeal construction and life process of the organism, be it a rose, a dolphin, or a human being. Failure to perform any one of these tasks would have relegated the DNA answer to the dump heap of failed theories.

Contrary to popular belief, Watson and Crick did not win the Nobel Prize for discovering DNA (that was done in 1869 by a now obscure biochemist named Friedrich Miescher); rather, they won it for explaining how DNA could carry out its various functions. Their great discovery in the early 1950s was how to fit together a puzzle that had been accumulating new pieces for 150 years.

The microscope propelled biology into the modern age; gross anatomy, which had been the staple of biology, was replaced by the study of the cell. Further refinements in optical technology led to the discovery that every cell had a nucleus and that every time the cell divided (a process called *mitosis*), the nucleus also divided. Better microscopes and staining procedures showed that the nucleus contained ribbons, or *chromosomes*, which divided during each cell division.

Although no one yet had any inkling of how it might work, these chromosomes seemed to be prime candidates for the "mystery factor" in the drama of renewal.

At the same time these investigations into the cell were happening (in the mid-nineteenth century), and without any knowledge of these microbiological developments, a monk named Gregor Mendel was experimenting in his garden by breeding pea plants. By carefully choosing characteristics which displayed themselves in a discrete all-or-none manner, and by working with breeding populations which were large enough to display convincing arithmetical trends, Mendel was able to prove that the hereditary material was composed of particles which could be transmitted independently of one another. In today's language, Mendel discovered that in the process of *meiosis,* the splitting of sex cells, characteristics were inherited as complete units and that certain traits predominated. The new science of heredity was born, and even though nobody understood the mechanism, it became possible to follow the path of different traits through many generations of plants and animals (usually the drosophila, or common fruit fly).

When Mendel's work was rediscovered at the turn of this century and combined with the knowledge about the cell's nucleus, a critical mass of information resulted. Around 1915, scientists concluded that the chromosome did not resemble a solid ribbonlike structure as much as it looked like a necklace composed of many beads. These beads, as many as 1,250 on a single human chromosome, were called *genes,* and the science of heredity gave way to the science of genetics. In the early 1940s, scientists at the Rockefeller Institute determined that genes were composed of nucleic acids rather than proteins. A few years later, another group demonstrated that not just any nucleic acid, but DNA, was the stuff of heredity. Finally, around 1950, the three components of DNA were analyzed: simple sugars known as deoxyribose, phosphate units, and four nitrogen bases—adenine (A), thymine (T), cytosine (C), and guanine (G).

All the puzzle pieces were now on the table. What Watson and Crick did was to explain how these components fit together to form a molecule which could perform the tasks necessary for heredity. The model they proposed was the famed double helix, or spiral staircase. The phosphates and sugars formed the twisted shape of the rails, and the bases formed the straight rungs of the ladder. Each rung was always formed of a specific base pair (e.g., A-T, G-C, etc.), and *the order of these rungs determined the exact traits* through their ability to specify which proteins were to be made. Since a single human DNA molecule may have as many as ten thousand rungs, the number of possible combinations is greater than the number of subatomic particles in the universe. As a conveyor of information, the DNA molecule has no peer.

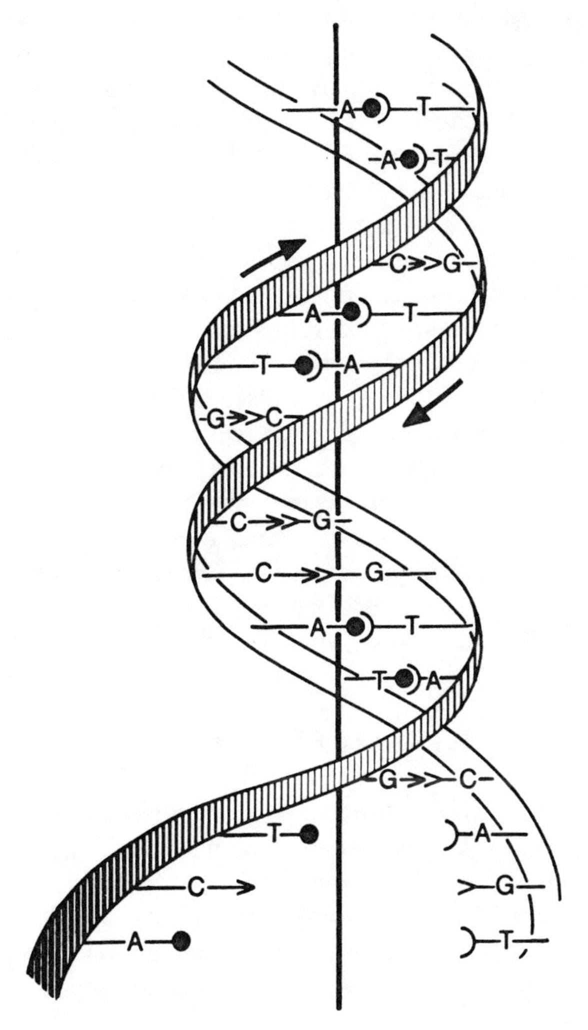

DNA—The World's Most Famous Molecule

The second requirement, an ability to self-reproduce, was solved exquisitely by the Watson-Crick structure. Prior to cell division, the DNA divides by unzipping itself. The two bases forming each rung snap apart, leaving a structure resembling a ladder sawed down the middle.

As the unzipping is taking place, free-floating nucleotides (a base attached to a sugar-phosphate backbone) start converging on both halves of the splitting ladder in order to repair them. The nucleotides can only couple with their specific base "mate," so the new DNA ladder is an exact replica of the original. Once the DNA has replaced itself, the cell divides and the process repeats itself.

The final requirement, the ability to supervise the construction of proteins, the building blocks of life, is met by a complicated mechanism that was not fully understood until well after the Watson-Crick work. Since proteins are made throughout the cell, while DNA resides only in the nucleus, an intermediary is needed. DNA partially unzips itself and uses its template to form a single strand known as *messenger RNA*. The messenger RNA then leaves the nucleus and goes to the cell's *ribosomes* —the subcellular protein-manufacturing organelles. The DNA also produces *transfer RNA*, which is used to lock the proper amino acids (the building blocks of proteins) into place according to the template provided by messenger RNA. Once the two RNAs are manufactured, the DNA zips itself up and the RNAs take over. Working together, they form a specific chain out of the twenty possible amino acids floating around in the cell. Like the base pairs in DNA, the sequence of amino acids determines the type of protein made. Instead of Aristotle's psychic powers, the mechanism of human renewal is a biological computer far more sophisticated than anything IBM is ever likely to produce.

CONVERSATIONAL TACTICS:

Twenty years ago, DNA was *the* topic of scientific conversation. The molecule appeared on the cover of national magazines, a gigantic model drew standing-room crowds at the 1962 Seattle World's Fair, and schoolchildren everywhere were stumbling over the word "deoxyribose." Today, the conversational action is usually directed toward some other topic, but DNA and the general study of genetics still present a number of interesting diversions. Try rolling the sides of your tongue up toward the roof of your mouth until it forms a long cylinder. The odds are seven in ten that you can perform this trick because you have the dominant gene for tongue-rolling. Three out of ten people received the recessive gene from both parents and can't perform this test. Fewer than one in one thousand have the gene which enables them to roll their tongue in the opposite direction. Other dominant genes which control the characteristics of the face and head include those for a widow's peak, a free earlobe, and a clockwise hair whorl.

Certain traits not only conform to the laws of dominant and recessive genes, they are also sex-linked. This means that one sex carries the gene for a certain trait while the other sex not only carries the gene but can also display the trait. As it turns out, women are almost always carriers, while males end up displayers of such nasty traits as color blindness and hemophilia.

RELATED TOPICS:

Altruism/Selfish Gene, Clones, Entropy, Information, Mutation, Origin of Life, Recombinant DNA, Sociobiology.

FURTHER READING:

The Body in Question, Jonathan Miller, Random House, 1978.
Chance and Necessity, Jacques Monod, Vintage Books, 1971.
The Double Helix, James D. Watson, NAL, 1968.
The Eighth Day of Creation, Horace Freeland Judson, Simon and Schuster, 1979.

ELECTRONIC SMOG

DEFINITION:

Electronic pollution caused chiefly by the proliferation of devices which use nonionizing radiation.

WHAT IT REALLY MEANS:

At the mention of radiation, most people think of A-bombs or nuclear wastes; unfortunately, there are now *other* sources of radiation to consider. The recent explosive growth in the use of nonionizing sources of electromagnetic radiation—such as thirty million CB radios, forty thousand miles of extra-high-voltage transmission lines, and widespread use of microwave ovens and microwave communication links—has increased the level of concern about the potential health hazards posed by these less energized forms of radiation. Some experts call it "electronic smog"—the toxic pollutant of the information age.

All electronic radiation may be defined as the propagation of energy through space by means of time-varying electric and magnetic fields. Different parts of the radiation spectrum are described by referring to their wavelength and frequency. X rays and gamma rays have the highest frequency (10^{19}Hz) and the shortest wavelength (3×10^{-8}cm). By comparison, visible light has a higher wavelength (3×10^{-5}cm) and lower frequency (around 10^{15}Hz). Microwaves have a frequency of approximately 10^{9}Hz and a wavelength of roughly 30 cm.

The specific numbers are not as important as the fact that they correspond to important physical properties which may be affecting the cells in your body as you read this. Perhaps the most important property is the difference between ionizing and nonionizing radiation—a distinction which splits the radiation spectrum into two distinct parts. Ionizing radiation refers to the ability of the waveform to change the internal structure of the atoms and molecules within the matter which is being irradiated.

X rays and gamma rays, which possess enough energy to dislodge orbital electrons from atoms, may create electrically charged, highly unstable, chemically reactive atoms called *ions.* Because of their high energy content, X rays and gamma rays inevitably damage the cells of living tissues, disrupt life processes, and cause genetic mutations. Microwaves, radio waves, and visible light have much less energy and cannot create ions; they are, therefore, forms of nonionizing radiation.

It is the other physical property of microwaves which makes them a valuable tool. Because they can be focused into compact, intense, highly directional beams, they have found wide application in communications. Most inter-city telephone communication is now accomplished through microwave relay systems. There are 250,000 microwave telephone and television signal relay towers in this country, and broadcasting satellites circle the earth, beaming back immensely powerful microwaves.

Because microwaves are reflected by electrical conductors and conveniently carried by waveguides, they are widely used in radar.

Microwaves also have thermal properties—matter which absorbs them is heated in the process—which have created a number of domestic uses. Medicine uses microwave diathermy machines; microwave cooking brings nonionizing radiation into the kitchen; industrial uses such as drying paint and curing resins make this radiation part of the work-place environment as well. Projected future uses include the wristwatch telephone, diagnostic uses in medicine, and radar-augmented braking for automobiles. Obviously, as the number of microwave devices and the power that they generate both increase, careful attention must be paid to any biological effects produced by nonionizing radiation.

The most striking fact about the biological effects of microwave radiation is how little we actually know about them. Although it is generally agreed that radiation is harmful in sufficiently large doses, whether the frequency is that of sound, light, or X ray, the only *uncontested* reports about the harmful effect of microwave exposure in humans are heat-related injuries to the skin and the occurrence of defects in the lens of the eye. Such thermal effects happen when the microwave energy is converted into heat inside the living organism.

There are, however, literally hundreds of *contested* reports which claim to demonstrate that microwave radiation can produce nonthermal effects. These studies have dealt with the effects microwaves might have on fetuses, on vulnerable types of tissues such as blood cells, on the immune system, and on the nervous system. Researchers in this country have tended to discount the reports of nonthermal effects and, therefore, to discount claims that low doses of microwave radiation might affect the nervous system. As a result, the recommended maximum occupational exposure permitted in the U.S. is a power density of ten milliwatts per square centimeter ($10mW/cm^2$).

In the USSR and other Eastern European countries, thermal effects were bypassed as the criteria for safety standards. Instead, nonthermal effects were the criteria used to set maximum permissible standards—reported physiological reactions, symptom complexes, and functional disorders among workers. Since the Russians regard these effects as functional changes which might become pathological processes, an important objective of their standards is to *prevent* possible organic disorders. The official standard set in the USSR is ten microwatts per square centimeter ($.01mW/cm^2$) for full-time work. This is a standard which is *one thousand times lower* than the standard for the U.S.

Given the conflicting medical evidence, should you consider buying a microwave oven or CB radio? Compared to the amount of microwave radiation already in the air, a little bit extra in your home or car probably won't do much damage. The sticky questions will arise with our first electronic smog alert: Will everyone be advised to sit at home without using their telephones or turning on their TVs? Will airplanes be grounded? How will the electronic smog be dissipated? How much of our health are we willing to trade for an efficiently run, convenient, smoothly communicating society?

CONVERSATIONAL TACTICS:

Unintended side effects—those little, and sometimes not so little, unplanned consequences of technology—are the stuff of engineers' nightmares. We've all heard stories, perhaps apocryphal, about the garage-door opener which set off all the fire alarms in the neighborhood, or the one about the automatic train doors which opened when the train was doing sixty. Next time the conversation happens upon technological glitches—those occasional jolts that tell us we're not always in control of our machines—you'll be prepared with some real tales to tell about electronic smog.

It turns out that those beepers doctors wear actually can sometimes jam sensitive life-sustaining equipment. In this scenario, your physician arrives just in time to kill you. Also on the medical front: One hospital in Washington, D.C., recently had to modify its electronic thermometers because FM radio signals were causing them to register 108 degrees when the patients' temperatures were normal. Then, in the hazardous-safety-devices department—transportation division—there is the story of the brakeless California school buses. They always worked perfectly in the garage but occasionally failed on the road. The cause was found to be electronic sensors buried beneath the road to control traffic flow; they were interfering with the buses' electronic braking systems. Finally, there's the peculiar case of *magnetic* smog. A researcher claims that the third rails of electric train systems (like BART in San Francisco or Metro in Washington) emit an ultra-low-frequency radiation which can confuse the brain and bring on drowsiness.

RELATED TOPICS:

Appropriate Technology, Energy Alternatives, Mutation, Technology Assessment.

FURTHER READING:

The Zapping of America, Paul Brodeur, Norton, 1977.
"The Radio Wave Syndrome," Michael Gold, *Science 80,* 1980.

ENDORPHINS

DEFINITION:

A group of polypeptides, found in the brain, which appear to mediate pain perceptions, moods, emotions, appetites, and memory functions.

WHAT IT REALLY MEANS:

Does the brain secrete thoughts as the stomach secretes digestive fluid? Do microscopic clockworks in our brains or molecular potions in our bloodstreams direct the course of our emotions? Is desire modulated by an invisible switchboard or an ultraminiature biocomputer? What neural currents or chemicals are involved in philosophy, insanity, or love? Everyone who seeks to understand human feelings must confront the same puzzle: How does physical tissue produce such metaphysical stuff as the manifestations of mind?

This much is known: That three-pound, wrinkled, gray organ packed into your cranium constitutes the most complex and functionally dense substance ever discovered. It will not be an easy task to map such a vast interior cosmos, but scientists have three crucial bits of knowledge to guide them. It is known that the *neuron* is the functional physiological unit of the brain, that the *nerve impulse* is the fundamental active force of brain function, and that nerve impulses are both *electrical* and *chemical* events.

The role of the peptides was revealed in the 1970s and they will undoubtedly be a fertile field of neurochemical research for years to come, but the true significance of endorphins is set against a background of research into nerve function that goes back more than fifty years. The electrochemical machinery of the nerve impulse has been investigated for at least that long, and it is at the level of the individual neuron that the psychic alchemy of endorphins takes place. The dual mechanism of brain processes is literally built into the architecture of the neuron.

A neuron looks more like an amoeba than a switching element or a little clockwork. It has a rather amorphous cell body that has numerous short extrusions; these small protuberances are the *dendrites,* and they handle incoming messages. A single, long, tubular extension of the cell body, known as the *axon,* serves as the conduit for outgoing messages. There is an important gap between each axon termination and the dendrites of other neurons: the connection between two nerve cells is known as a *synapse,* and the gap is known as the *synaptic cleft.*

The electrochemical nature of brain function is determined by a clever mechanical arrangement in the membrane of the cell. The axon is vastly more sophisticated than a simple wire for the conduction of pulses, and the nerve impulse is more like a wave than a spark. The membrane of the axon is penetrated by small gates and valves which are just the right size to let atoms of specific sizes in or out. The inside of the nerve cell is electrically negative with respect to the outside of the membrane because positively charged sodium ions are constantly pumped out of the sodium-sized gates by an internal mechanism of the cell. When enough signals from other nerve cells come in through the dendrites, the gates at the base of the axon begin to let sodium ions back into the cell, temporarily making the interior of the membrane less negative (or more positive) with respect to the outside. This "depolarization" travels like a wave down the axon, stimulating the sodium gates to open all along the route. When the nerve impulse passes a point on the axon, the sodium gates close again to the outside, while other gates let positive potassium ions leave the cell to restore the polarity.

At different parts of this cycle, the nerve impulse is mediated by either electrical or chemical events, or both. As the wave of electronic depolarization reaches the end of the axon, chemistry takes over again. At the end of the axons are packages of chemical messengers known as *neurotransmitters.* The arrival of the impulse sends the chemical message across the synaptic cleft to the dendrites belonging to another nerve cell. The neurotransmitters chemically trigger the dendrites, which in turn initiate a depolarization wave down the axon of the "receiver" neuron. In less than one ten-thousandth of a second, the dendrites release the neurotransmitters, which are either broken down into components by special enzymes floating in the synaptic cleft or find their way back to their originating neuron and are degraded within the axon.

Neurotransmitters can either excite or inhibit nerve impulses, desensitize or sensitize dendrites, and alter the action of other neurotransmitters. An important branch of neurochemistry concentrates on the connection between these chemicals and observable brain functions. Cocaine, for example, appears to create euphoria by blocking the chemical pump that returns one kind of neurotransmitter to the originating cell. Because they remain in the cleft, where there are no enzymes to deactivate them, these molecular messengers keep delivering the same messages over and over again, tickling the receptors on target neurons. Similar mechanisms seem to be involved in other mood-altering drugs.

The ability of drugs like morphine to alter the mood of human subjects and to block pain perceptions has been one of the most fruitful questions of psychopharmacology. Avram Goldstein of Stanford, one of the pioneers in the endorphin investigation, began the quest for a whole new class of nerve chemicals *beyond* neurotransmitters by asking a simple question: Why should a chemical extracted from poppies have such a strong affinity for parts of the human brain? It seemed likely that morphine must chemically resemble a substance that exists naturally in certain brain cells, even though no such substance was then known. In 1975, two breakthroughs in this search signaled the beginning of the "peptide revolution."

In December of 1975, John Hughes and Hans W. Kosterlitz of the University of Aberdeen, Scotland, extracted an opiatelike substance from pig brains. They named this short molecule (a chain of five amino acids) *enkephalin,* for "in the head." Six months before the discovery in Scotland was reported, Goldstein found a larger molecule in the pituitary gland, the organ which commands the hormone system. When this substance, named *endorphin* for "the morphine within," was more closely analyzed it was found that enkephalin was identical to the last five units on the endorphin chain. Other substances with different functions were soon discovered, and it quickly became clear that the brain does manufacture hormones to regulate its drives, moods, appetites, and emotions.

The new class of neurotransmitters were *peptides*—very short proteins consisting of chains of amino acids strung together like beads on a chain. While the normal neurotransmitters are very small molecules, with molecular weights on the order of one hundred, enkephalin is around one thousand, and peptides are around ten thousand. Some scientists are already calling the peptides "neuromodulators" because they seem to modify the responses of neurons to other neurotransmitters. There appears to be one large "mother" peptide which is chopped up into little pieces as needed; one of the component chunks modifies the activity of pain-mediating cells; others can induce tranquillity or rage, hunger or satiation.

CONVERSATIONAL TACTICS:

The most intriguing aspect of endorphins is not how they were discovered, but what they can do. Everything from the ultimate painkiller to the first true aphrodisiac may soon be derived from peptide research. The rate of discovery of new brain peptides has moved far faster than the rate of application. Although years of basic research and years of careful pharmaceutical screening lie ahead, it can be safely predicted that a few of the new peptides in the list below will lead to a pharmaceutical revolution in the 1990s that could dwarf even the massive social effects of the birth-control pill in the 1960s.

LRF and *LHRF* are the shorthand designations for the tongue-bending names of peptide hormones which show promise as unisex birth-control substances and possibly as aphrodisiacs.

Beta-endorphin is forty-eight times as powerful as morphine; *dynorphin* is two hundred times as powerful as morphine; and at the very opposite end of the pain spectrum, *bradykinin* has been called "the most painful substance known"—even the tiniest amounts can cause intense pain when injected.

Vasopressin and *MSH/ACTH* 4-10 may be true intelligence-amplifying drugs; they have been found to improve visual retention, help concentration, and stimulate memory functions.

Cholecystokinin may be the substance that tells the brain to send out the signal to stop eating. *Bombesin* may be involved in determining whether incoming calories are metabolized or turned into fat. The true diet pill may be the result of either or both of these lines of research.

Gamma-compound appears to be a natural receptor for Valium, in the same way that enkephalin is a receptor for morphine. The search is on for other peptides which either relieve or trigger anxiety.

RELATED TOPICS:

Acupuncture, Consciousness, Intelligence Enhancement, Placebo, Reproductive Technology.

FURTHER READING:

"Chemical Feelings," Joel Gurin, *Science 80,* January, 1980.
Frontiers of Science, National Science Foundation, 1977.

ENERGY ALTERNATIVES

DEFINITION:

Because alternate energy sources involve economic, political, and social concepts as well as scientific and technological ideas, there can be no single, precise definition. In fact, much of the debate concerning energy is a dispute about the relative merits of the scientific data and supporting technological information.

WHAT IT REALLY MEANS:

Personal preferences regarding energy policy have become a kind of political code. One's position on nuclear energy is linked to one's opinions about foreign aid; attitudes toward synthetic fuels shape other attitudes regarding farm price parity; support for solar technology has become entwined with tax reform. Energy is one of the key issues of the 1980s.

It was not always such a controversial field. Prior to 1973 and the Arab oil embargo, energy was plentiful and cheap. There was no need to cast about for alternative energy sources; we had, or could import, all the oil and natural gas we needed. About the only energy-related topic of conversation was the shape of the free glassware the gas station was giving away this week. That situation changed with dizzying rapidity. The price of gasoline doubled and redoubled. The price of a barrel of OPEC crude increased eightfold. The U.S. automobile industry went into shock, there was an increase in migration to the Sun Belt, and in some places it became necessary to make an appointment to "fill 'er up." If it wasn't always clear that the world was running out of oil, it was at least clear that the era of cheap petroleum fuel was finished. The alternative energy business was born. But just what was it really selling?

The answer depended on the salesman. Successive administrations, worried about the political consequences of energy dependence, sought to secure an adequate domestic supply of energy. To this dominant faction, the key to alternative energy supplies was increasing the use of coal (the U.S. has been called the Saudi Arabia of coal), building more nuclear power plants, and developing synthetic fuels. Dissident ecological groups would have none of this; they perceived a crucial difference between energy and fuel. Energy, from the Greek *ergon,* meaning "work," is simply the capacity to do work. Fuel is a substance which must be expended to produce energy.

From the ecological standpoint, the world might be suffering from a fuel shortage, but not from an energy shortage. To the advocates of this view, alternative energy sources provide an opportunity to stop using polluting fossil fuels and potential radioactive polluters and to shift instead to cleaner sources of energy: solar, hydroelectric, geothermal, etc. A third group, closer in spirit to the environmentalists, advocates using only renewable sources of energy; they propose fuel sources such as "BTU bushes," "gasoline trees," and "kelp farms."

Ultimately, what emerged from these opposing lobbies was a fusillade of claims, arguments, schemes, and statistics which still haven't been sorted out. While a discussion of the relative economic claims regarding shale oil recovery is beyond the scope of this book, it is important for every citizen to understand the difference between a photovoltaic cell and a breeder reactor. To that end, *Talking Tech* presents its alternative energy lexicon—seven technologies to rival the "seven sisters" of big oil.

Biomass Conversion

Technically, *biomass* refers to any organic object. However, in the energy sphere, biomass refers to special crops which can be cultivated for their speed in storing solar energy. Biomass conversion refers to the various techniques used to obtain the chemical energy contained in these crops.

Anyone who ever sat around a campfire has engaged in the most fundamental form of biomass conversion. Wood was, and in many areas still is, humanity's first fuel crop. Even though we know how to farm trees, no one is suggesting wood as a practical solution to the global energy crisis. (Remember, England got into coal because of an energy crisis—all the forests were cut down for fuel!) Instead, biomass conversion is presently concentrating on liquid fuels as replacements for oil. Gasohol is the best known of the biomass fuels; over two thousand U.S. gas stations are now selling a mixture of 90 percent gasoline and 10 percent grain alcohol.

Fuel alcohol can be distilled from any grain crop (sweet sorghum is the most efficient, with a yield of almost four hundred gallons of alcohol per acre), and gasohol has a higher octane rating than pure gasoline. Opponents of gasohol claim that a world on the brink of mass starvation can't afford to channel important food crops into fuel crops. Will the car of the future literally take its fuel out of the mouths of babes? Proponents counter that gasohol does not have to be made from grain; it can also be made from agricultural wastes or from sea kelp, the fastest-growing plant known to agri-science.

Ultimately, all energy production ideas must be judged by the criterion of net gain—energy produced minus energy expended in production. Distilling crops to produce energy consumes substantial energy, so the gain is not great. Solar distilleries improve the gain, but direct extraction of a liquid fuel is a much more potent idea. Dr. Melvin Calvin, who won the Nobel prize for explaining how plants convert sunlight into energy, believes that there is a class of desert plants, such as the jojoba, which could be tapped for their hydrocarbon sap the way maple trees are "sugared." He envisions vast fuel plantations in the Southwest where these "gasoline trees" could be grown without competing against food crops. If the biomass enthusiasts are correct, "Put a cactus in your tank" may be an advertising slogan of the future.

Coal

Can the fuel of the past become the fuel of the future? Coal is abundant: One reliable estimate puts the U.S. recoverable deposits equal to four times the energy potential of all Arabian oil fields. Coal is not easy to mine, transport, or burn cleanly, however. The complications of using more coal can be seen in this table comparing the costs and benefits of strip mining:

Benefits	*Costs*
1. More usable energy	1. Water pollution
2. State/federal extraction royalties	2. Thermal pollution
3. Increased employment	3. Acid mine drainage
4. National security	4. Air pollution
5. Export goods	5. Diminished grazing, forest, and agricultural land.

Coal's future will be decided on political grounds, but a few technological concepts will be bandied about when the arguments get going.

Probably the most widely recognized problem associated with the use of coal is air pollution—coal soot actually turned industrial Britain black in the time of Dickens. This potentially lethal by-product is being combated by smokestack scrubbers—devices which attempt to cleanse the emissions as they leave the stack. However, scrubbing doesn't remove the nitrogen oxides or several other pollutants. In its place, fluidized bed combustion is being developed. This system uses a bed of ash plus continuously-fed crushed coal, which is levitated by streams of air from below. The coal burns at a relatively low temperature in the bed, minimizing the production of nitrogen oxides.

One major obstacle to increased coal use is transportation. The nation simply doesn't have enough rolling stock or good miles of track to move all the coal that this method's proponents call for. One solution to this problem is the development of coal slurry pipelines—coal is crushed and mixed with water and transported through pipelines. While this method is cheaper than building railroads, it is heavily dependent on enormous amounts of water, a rare commodity itself in the coal-rich western plains. Nobody doubts that the coal is there. Everyone argues about how to mine it, move it, and make it into energy.

Geothermal Energy

No one who has witnessed Yellowstone's Old Faithful or who remembers the fate of Washington's Mount St. Helens can have much doubt that the earth is hot stuff. That is literally what geothermal means—*earth heat*—and it's in stark contrast to the old cliche about a cold, cruel world. Geologists now know that only the outer crust of the earth is cold; most of the earth's 260 billion cubic miles of rock are above the melting point of 2200 degrees Fahrenheit. Geothermal engineers seek ways to make that heat usable here on the surface.

There are currently five major approaches, depending on the source and configuration of the heat. The easiest to harness is dry steam, which can be used directly to drive a turbine to produce electricity. Unfortunately, there are not many sources of dry steam in the U.S. The only large commercial dry steam power plant is The Geysers, which provides about half of San Francisco's electrical requirements.

A more promising source of energy is subterranean hot water under pressure. This water, trapped by cap rock, is in a caldron with temperatures in excess of 300 degrees Fahrenheit. By drilling wells into the caldron, hot water can be brought to the surface in the form of steam to drive turbines. The entire southwest region of the U.S. is honeycombed by these hot water pockets, and work is under way to determine the economic feasibility of large-scale geothermal power plants.

A third possibility uses water which is less than 300 degrees, the minimum temperature needed for efficient production of electricity. Iceland has long used geothermal hot water to heat its capital city, and Idaho is considering using Boise's geothermal hot water to heat state buildings. Eventually, by using the water to heat low-boiling-point liquids such as Freon, electricity may be produced through a vapor-turbine-cycle technique.

Two additional sources of geothermal energy look promising but are still years away from practical development. The first source uses "geopressurized zones," which are large hot water reservoirs trapped below thick sedimentary deposits. Found in Texas and Louisiana, these areas contain huge amounts of methane—the chief ingredient in natural gas. The other future source is known as "hot dry rock"—heated rock near the earth's surface that lacks water to carry away the heat. By forcing cold water down a well, circulating it through the rock's fissures, and bringing it back through a second well, engineers hope to generate steam. Geothermal energy has many advocates because it is relatively clean, fuel-free, and abundant. If the geothermal partisans are proven correct, hot rocks may mean something much more valuable than "stolen jewels" in the near future.

Nuclear Power

If you can remember backyard fallout shelters, then you probably remember the optimistic claims made for nuclear power—electricity so cheap it wouldn't need to be metered, more energy in a cup of uranium ore than in ten barrels of oil, and the hopefully named "Atoms for Peace" program. Even though we can't tell you how to vote on a nuclear plant site initiative or recommend which petition to sign, we can tell you something about the technology behind nuclear power.

All the energy produced by nuclear power in this country now comes from fission reactors. As discussed under Fission and Meltdown, these reactors use the heat generated by splitting atoms, usually uranium 235, to heat water into steam to drive turbines. Although they are an alternative to fossil fuels, fission reactors look less and less like the alternative of the future because of negative public perceptions and a growing uranium scarcity. Already, two new nuclear technologies are being developed to replace the present reactors.

The first of these, liquid metal fast breeder reactors (LMFBR), can convert nonfissile uranium 238 into fissile plutonium 239. This means that more than 70 percent of the energy potential of the uranium can be tapped, as compared to the less than 1 percent available in conventional fission reactors. It also means that LMFBRs could operate for millions of years on the known uranium reserves in the U.S., in contrast to the one hundred to two hundred years' worth of uranium reserves under the current system. That's the good news.

The bad news is that breeder reactors require reprocessing and recycling of fuel in order to achieve their resource-conserving potential. Many have voiced concern that, in addition to being technologically tricky and medically dangerous, the plutonium could be illegally diverted to terrorists or irresponsible nations. A plutonium economy, say its critics, could turn "Atoms for Peace" into "Atoms for Terrorism." For the time being, the U.S. has deemphasized breeder technology.

The second alternative, nuclear fusion reactors, has neither the radioactive waste problem nor the strong weapons connection associated with conventional fission and breeder reactors. As discussed more fully under Fusion, these reactors produce a gas of deuterium and tritium ions ten times hottr than the core of the sun and confine this plasma in a magnetic field long enough to produce helium, neutrons, and energy. Although fusion reactors seem to have the most promise for the long term, many technological obstacles remain, and it will probably be well into the twenty-first century before any U.S. energy is produced by fusion.

SOLAR ENERGY

Virtually all of the energy available on earth is derived from the sun. The fossil fuels we dig out of the ground were once living plants, millions of years ago; the solar energy those plants captured through photosynthesis is still locked into the chemical structure of their remains. Biomass is a way to organically convert the sun's energy without waiting that long. Wind is caused by differentials in the atmosphere caused by the sun's heat, and hydroelectric power is possible because solar heat drives the earth's rain cycle. All of this demonstrates what Leonardo da Vinci deduced centuries ago: "The heat of the universe is produced by the sun."

Today's solar scientists are trying to use solar radiation directly in three ways: heating and cooling, solar-thermal electric power, and electric production from solar cells. Individual home use of the sun's energy for heating is a major growth industry. Typical "passive" solar-heating systems collect the sun's heat with rooftop arrays of piping and flat metal sheets painted black to absorb as much radiation as possible; such collectors are encased in glass or plastic and angled southward to catch maximum sunshine. Air or water in the piping distributes the heat through ducts or stores it in an insulated water tank or a bin of rocks. If you are considering converting to solar heat, take a close look at the map of the U.S. Draw a line from San Francisco to Bismarck, North Dakota, then south through St. Louis, Atlanta, and finally to Miami. If you live south and west of that line, your home probably averages in excess of 2800 hours of annual sunshine, so consider yourself a denizen of the Sun Belt.

If passive solar heating is a technology for the individual home owner, then solar-thermal electric power is a possible technology for the power utilities. Steam boilers used in generating electricity require temperatures of about 1000 degrees Fahrenheit. By comparison, a conventional flat-plate solar collector seldom gets above 200 degrees. Putting sunshine to work to provide electricity will require devices which concentrate the sun's rays. While there are many different designs, they all share common features: a huge series of mirrors, or heliostats, which are mounted so that they track the sun, and a water boiler mounted on a tower to absorb the reflected rays. History records that Archimedes repelled the Romans by burning their fleet with concentrated solar rays. If history repeats itself, Americans may repel the oil sheiks by converting the sun's radiation into electricity.

A more elegant way to convert the sun's power into electricity is through solar, or photovoltaic, cells. A typical solar cell is a thin wafer about the size of a silver dollar. It is sliced from an ingot of pure silicon crystal, which contains a minuscule amount of impurity that allows the crystal to conduct positive electric charges. A different impurity is diffused into the top of the wafer, allowing that section to conduct negative charges. In essence, the two sections behave like the oppositely charged poles of an ordinary car battery. When photons of light strike the cell, they stimulate the flow of positive and negative charges—electricity.

Energy Alternatives—A Solar-Powered Reading Hat

These cells are still too expensive for mass use, but scientists predict that the day may come when solar cells are delivered to a house like rolls of roofing paper to be tacked on and plugged into the wiring—making the home its own power station. In general, photovoltaic and other solar energy ideas seem to be in the same stage nuclear energy was thirty years ago—there is much promise, but many technological problems have yet to be solved. It would be wise not to oversell it.

Synthetic Fuels

A major problem hindering the increased use of coal is its physical form. Most homes are no longer outfitted to use coal as a heating source; you can't pump it into a gas tank; and coal-fired steam engines gave way to diesels long ago. But all fossil fuels contain similar atomic structures, and it should be possible, in principle, to convert coal to a gaseous or liquid form. In fact, coal gasification has been around for over a century. Such synthetic gas, however, has only 10–15 percent of the heat content of natural gas.

Recent advances, however, add a separate step that boosts the coal gas's methane content to the same level as that found in commercial natural gas. The HYGAS technique, which relies on a chemical reaction between coal and hydrogen, first provides a low-heat, low-energy gas; then the gas is cooled, rinsed, chemically scrubbed, and passed through a nickel catalyst to impart additional methane. The result is SNG, substitute natural gas so pure it can flow directly into the nation's natural-gas pipelines.

Coal liquefaction was used by Nazi Germany during World War II in order to manufacture fuel to run the war machine. They employed a technology known as the Fischer-Tropsch process. First, low-BTU synthetic gas is produced from coal, then it is converted to a variety of liquid fuels. Modern systems are much more efficient and promise to yield 2.5 to 3 barrels of liquid per ton of coal, but any use of coal is constrained by the problems already mentioned. Syn-fuels also require tremendous amounts of water and are notoriously toxic. Before the much ballyhooed syn-fuels research program can succeed, many technological obstacles will have to be overcome.

Wind Power

Ever since Don Quixote, tilting at windmills has meant challenging an impossible problem. Modern scientists believe that wind-driven generators might be just the right challenge for the energy program. Of course, wind power isn't a new idea. What is new is the idea of using it on a grand scale. Windmills ground the grain in medieval Europe, and farmers on the Great Plains used windmills to draw their water. Now there are plans for huge phalanxes of wind-driven generators to replace fossil-fuel power plants.

While you're waiting for such plants, you might consider the economics of building your own windmill. The best places in the U.S. for strong winds are the Northeast coast, the west coast, and the western Great Plains. Since wind power increases with the cube of velocity (a 20 mph wind is 8 times as powerful as a 10 mph wind), a little more wind can mean a lot more energy. Scientists have found that flying special kites is the most accurate way to "prospect" the wind. Windmills now seem romantic—vestiges of a time long past. If they end up supplying 5 to 10 percent of our energy by the year 2000, as some have predicted, we might have to look for a new nostalgic symbol. Perhaps the oil barrel will do.

CONVERSATIONAL TACTICS:

Budding technologies always produce many bizarre, Rube Goldberg ideas. Remember all the crazy attempts at flight memorialized by Movietone News? In addition to the big-seven alternative-energy technologies presented above, there are many more speculative ideas floating around today. Who knows just which ideas might succeed? Without passing judgment on their viability, *Talking Tech* offers this list of "farther out" energy technologies.

Black Holes

Societies of the distant future may be able to "farm" black holes as an energy source. Simply using the gravity waves associated with a black hole could provide energy at six times the efficiency of the H bomb. If two black holes could be fused, the energy potential would be sixty-five times as efficient as the H-bomb. There is a serious question, however, about whether we would ever be able to safely approach such energy-producing dynamos without getting sucked into their gravity wells.

Cow Burps

From the sublime to the ridiculous: A few years ago, some agronomists calculated the amount of methane gas burped up by cows and concluded

Energy Alternatives—The Cow-vrolet

that if we could capture it, we could supply a good percentage of our national needs. As far as we know, no one has yet patented a cow-burp catcher. Carl Sagan, famed exobiologist, has pointed out that beings on another galaxy might not be able to detect intelligent life on earth, but they would be able to surmise the presence of methane burpers from spectral absorption lines! SRI International, a well-known think tank, included the growing volume of methane emissions into the atmosphere in its list of potential ecological problems facing the next generation.

Harnessing the Gulf Stream

The Gulf Stream is a hydraulic dynamo. Two respectable scientists, Peter Lissaman and Bill Mouton, now think they have a way of harnessing that energy to make electricity. They envision a series of 230 turbines, each larger than three 747 jets, strategically placed in the Gulf Stream off Florida. This arrangement might be able to supply the energy equivalent of ten nuclear plants. The final cost is estimated at ten billion dollars.

OTEC

Ocean thermal energy conversion (OTEC) is another approach to retrieving energy from the sea, and it is also an approach which is supported by government grants and industry feasibility studies. OTEC energy converters rely on the difference between the temperatures of ocean water on the surface and that 1500 feet below the surface. Heat from the warm surface waters would vaporize a working fluid such as ammonia, and the vapor would drive a low-pressure turbine. The ammonia would then be recondensed to a liquid form by the cold water pumped up from the depths.

Solar Satellites

Our artificial satellites have always used solar cells to power themselves while in orbit. The idea now is to bring that power down to earth. Satellites with a wing span of seven miles would be placed in geosynchronous (stationary) orbit above a large city. The direct-current electricity produced by the satellite's photovoltaic cells would be converted into microwaves and beamed to a huge ground-based receiving antenna. There the current would be converted into alternating current and distributed for use. One "glitch" is the expense involved in putting up the satellite. Professor Gerard O'Neill, father of the space colonization movement, wants to have space colonists build the orbiting stations, using materials from the moon! Another glitch is the weapons potential of such orbital power stations. Could those energy beams be converted for destructive purposes?

Weight Power

Wayne P. Le Van wants to generate power by harnessing the weight of cars and people. He proposes installing gratelike "hit plates" on busy sidewalks and streets. The weight would depress the plate slightly, forcing a noncompressible fluid underneath to flow through hydraulic hoses to a mechanism that would turn a generator. If Le Van is right, fat people and big, heavy cars may come back into fashion.

RELATED TOPICS:

Appropriate Technology, Fission, Fusion, Laser, Meltdown, Photosynthesis, Technology Assessment.

FURTHER READING:

Energy in Transition, 1985–2010, National Academy of Sciences, W. H. Freeman, 1979.

Energy Future, R. Stubaugh and D. Yergin, Random House, 1979.

"Energy," *National Geographic,* February, 1981.

ENTROPY

DEFINITION:

A measure of the randomness, chaos, or disorder in a statistical or mechanical system; also, a measure of the energy unavailable for use in a thermodynamic system.

WHAT IT REALLY MEANS:

You can't brew today's tea with the hot water from yesterday's kettle. The energy your stove originally put into the kettle is now dissipated into the air. When you mix hot water and cold water you get lukewarm water, which will eventually grow cold but which will never get hotter unless energy is added. The death sentence of the entire universe is written in each increment of lost energy. The same energy economy which governs the brewing of tea also rules the energy of atoms and galaxies.

Although they didn't call it entropy, the great religious figures of history were conscious of this tendency toward chaos and decay. Twenty-five hundred years ago, as he was dying, the one they called the Buddha reminded his disciples that "all compound things are subject to decay." Half a millennium later, on the other side of the world, Jesus exhorted: "Lay not up for yourself treasures on earth, where moth and rust doth corrupt. . . ." (Matthew 6:19). This message of physical impermanence, stated in its most metaphysical form, transformed the spiritual lives of millions of people for thousands of years.

More recently, and in secular guise, the experimental examination of the same concept has transformed the science and technology of the modern era. When the Industrial Revolution was cranking up to full speed in the mid-nineteenth century, the scientific study of engines was a matter of great practical interest. As physicists began to measure the efficiency of energy-converting machines such as steam engines, they stumbled onto a strange quirk in the way the universe works: No engine can ever be totally efficient, because the tendency is for entropy to constantly increase—the universe itself is continuously and irreversibly losing usable energy. This means that no machine can ever produce more energy than is put into it. It also means that all the available energy everywhere will one day be useless, frozen, and locked up by entropy—*the heat death of the universe.*

These far-ranging statements were given scientific precision by *thermodynamics*—the study of heat exchange. The first law of thermodynamics states that the energy of a closed system (one which receives no additional energy) remains constant. Within such a system, energy can never be created nor destroyed. The distilled philosophical implication of this physical principle is that "you can't win" on the cosmic scale. The second law of thermodynamics, which involves entropy, states that the energy in every closed system becomes less available for use as time passes. This might be philosophically translated as "you can't break even."

One of the earliest scientific definitions of entropy was proposed by Rudolf Clausius in 1865: Heat can flow from a hot body to a cold body, but never from a cold body to a hot body, and once the two bodies reach the same temperature, the heat flow stops, permitting no more work to be done. When it was discovered that heat was a measure of the average energy level of a population of molecules, entropy's definition was generalized: Closed systems tend toward their most probable, random, disordered state. Clocks fall apart, but pieces of metal don't fall together to spontaneously form a clock.

Today, entropy is precisely described by statistical mechanics. It is no longer simply the measure of the technological efficiency of a mechanical system; it is a general principle, as true for the stars as it is for the steam kettle. This principle has proved to be one of the most adaptable ideas ever developed, finding its way into most of the sciences, including biology, chemistry, and mathematics.

Entropy is also part of a profound paradox: If it is true that the universe will never again be as ordered, energetic, and complex as it is right now, then we might well ask, "What are we doing here talking about it? Isn't life and consciousness as unlikely as a clock assembling itself from its components?"

Biological evolution is the story of increasingly improbable, ever more highly ordered, progressively more complex systems. How do humans, and all other life forms, manage to swim against the entropic stream? And how did a bunch of chaotically disorganized elements manage to get together and develop into meaningful patterns in the first place? The simplest answer to the first question is that terrestrial life continues to defy the cosmic energy tide, courtesy of our sun's entropy. As the sun radiates energy—a finite process which will one day cease—photosynthesizing cells in plants capture some of the dissipating solar energy and convert it into chemical energy, which higher life forms use. As long as the sun keeps shining, earth is not a closed system.

The second question—how life could have originated—concerns the nature of chemical reactions. When two substances, each containing a specific amount of chemical energy, are combined, they form a single substance with less chemical energy. Entropy is expressed as the heat lost during the reaction and the reduced chemical reactivity of the product. Since all chemical reactions tend toward equilibrium in this way, the second question is a puzzle: How can the heavily imbalanced reactions needed for the genesis and evolution of living organisms overcome the entropy barrier?

A cryptic answer to this puzzle was suggested in 1945 by Erwin Schrödinger, the distinguished quantum physicist. In a justly famous lecture titled "What Is Life?" he remarked that "living organisms eat negative entropy." In other words, life is able to become progressively more ordered (negentropic) in an increasingly disordered (entropic) universe by *feeding* on the disordered environment. This made quite a bit of sense to Francis Crick, who subsequently switched his major subject from physics to biology and ended up, along with James Watson, winning the Nobel prize for explaining the principle of life's negentropic engine—the informational structure of the DNA molecule. On the cosmic battlefield, information and structure are the mortal enemies of entropy because they introduce order.

Schrödinger's lecture also made sense to Alan Turing, a brilliant logician and mathematician. Turing's work into how patterns could arise in space and time spontaneously from undifferentiated matter led to the discovery of the oscillating chemical reactions which make the origin of life possible. He built mathematical models of the entropic properties of chemical reactions, which a Belgian chemist named Ilya Prigogine used to demonstrate the existence of rare chemical reactions; these unlikely molecular occurrences allow the second law of thermodynamics to remain valid for the universe while permitting entropy to fail in local conditions, such as those which exist on earth.

At a distance far from equilibrium, just the right kind of random event on a very small scale can trigger much larger waves and cascades of reaction, amplifying these chance fluctuations in the entropy flow. The amplified fluctuations can lead to configurations of a more complex order, structures capable of interacting with the environment, of locally reversing entropy by feeding on complexity and eliminating wastes. Prigogine won the Nobel for his model of these "dissipative structures"—open systems that fight the downstream flow of energy by constantly exchanging energy with the environment. The key to this sophisticated energy-exchange policy is the information encoded in the DNA molecule.

Communications engineers were also infected by a similar idea when Claude Shannon, studying the transmission of messages, noted that the probability aspect of entropy makes negative entropy mathematically equivalent to information! The more information in a system, the less entropy. We life forms owe our improbable existence to the molecular equivalent of knowledge—useful information at the subcellular level.

CONVERSATIONAL TACTICS:

Talking Tech offers an anecdote and a simple demonstration to assist any conversation having to do with entropy. The anecdote concerns James Clerk Maxwell, one of the nineteenth-century scientists whose work intersected with the idea of entropy. His other work led to the quantum and relativity revolutions, but his contribution to thermodynamics and information theory is that delightful hypothetical creature known as "Maxwell's demon." Consider a container with a barrier small enough to pass only one molecule at a time. On one side of the barrier is a very energetic, hot gas; on the other side is a less energetic, cooler gas.

According to the second law of thermodynamics, the hot molecules should migrate across the barrier, losing energy in collisions with slower molecules, until both sides of the container are the same temperature. But what (Maxwell wondered) would be the effect of placing a little demon at the molecular gate, an imp who would let only the warmest molecules of the cold population pass through one way and the coolest molecules of the hot population pass the other way? The hot side would get hotter and the cold side would get colder, reversing the flow of entropy! How could this be accomplished? When information theory was born nearly a century later, it became clear that Maxwell's demon possessed one special talent that enabled it to perform its paradox—the demon knew when one kind of molecule or the other was approaching the barrier. The demon possessed *information.*

The demonstration of entropy is a simpler story. Simply dump a teaspoonful of maraschino cherry juice, wine, or food coloring into a glass of water. Tell your conversational companion that this act demonstrates the one-way flow of entropy: "It was easy for me to put it in—now you try to get the dye out!"

RELATED TOPICS:

DNA, Information, Mathematical Modeling, Origin of Life, Photosynthesis, Stellar Death, Uncertainty Principle.

FURTHER READING:

General System Theory, Ludwig von Bertalanffy, Braziller, 1968.
The Eighth Day of Creation, Horace Freeland Judson, Simon and Schuster, 1979.

EXOBIOLOGY

DEFINITION:

The study of living organisms which originate and reside outside the earth's environment.

WHAT IT REALLY MEANS:

Taken literally, exobiology may well be a subject without any subject matter. In fact, if it turns out that there are no life forms elsewhere in the universe, we may be hard pressed to differentiate exobiology from unicornology. Why, then, has the government spent millions of dollars on exobiology research, and how has that money been spent?

One way to resolve this paradox is to broaden the definition of exobiology to include the study of environmental conditions and the possible biochemical and evolutionary pathways which might lead to life beyond earth. As such, exobiology experiments may be conducted in laboratories, where scientists search for the key to life on this planet as well as its possible existence on extraterrestrial bodies. This type of earthbound research, however, is not the core activity of exobiology and is discussed instead in the chapter Origins of Life.

The real key to understanding the excitement over exobiology is found in two words: *payoffs* and *procedures.* It is simply more important to discover a slime mold on Europa (a moon of Jupiter) or an intelligent being in the Andromeda galaxy than it is to discover a horse with a horn protruding from its forehead. Therefore, even if the odds of finding space creatures are no better than those of finding a unicorn, the potential payoff is great, and the search goes on.

However, even though the end goal of exobiology is important, it is its procedures which qualify it as a science. The major work of exobiologists is generally carried out through one of two experimental methods. First, within our own solar system, we launch probes to take pictures as well as to actually study the atmosphere and soils of other planets. In this way, we are searching for direct physical evidence of other life forms. But our solar system is only one out of the trillions in the universe, and all the rest, because of their distance, cannot be directly assayed. Therefore, the second method is to send, and attempt to receive, messages from other beings. This implies that these other beings are also intelligent enough to send or receive messages, something we earthlings have been able to do for only a few decades. This branch of exobiology is actively searching for extraterrestrial intelligence (ETI).

The most famous attempts to discover other life in our solar system were the Viking missions to Mars. The Viking landers were actually little robot-run laboratories, dropped on the martian surface and programmed to conduct a series of five types of experiments to see if there was evidence of any martians. First, cameras scanned the landscape for any signs of organisms. Second, the soil was analyzed in order to detect any signs of organic molecules containing carbon. Third, the soil was further analyzed to determine if anything in it was converting carbon dioxide into chemical compounds. Fourth, water and nutrients were mixed with soil and studied for the release of oxygen or carbon dioxide. Finally, the soil was tested to see if anything in it might consume chemicals as terrestrial animals do. Although some of the experiments failed to give clear-cut answers, the general conclusion was that Viking did *not* detect life on Mars.

Partially because space missions are so expensive, and partially because they limit us to searches in our own solar system, the bulk of exobiology research is carried out through radio astronomy. Because we are "new kids on the block," most of that work, as described in the chapter Radio Astronomy, involves *listening* for extraterrestrial messages. On occasion, however, we have tried to make our presence known to anyone (anything) who (which) might be listening. The best-known message was sent in 1974 to a collection of stars known as M13. You shouldn't be miffed, though, if we haven't received a reply. Since M13 is 24,000 light years away, it will be 48,000 years before we receive a reply—assuming that someone is listening.

What does one say in a message that long-lived? How does one keep the news from going stale? First of all, the message is not in English or any terrestrial language, but in a mathematical code. Exobiologist Carl Sagan, writing in *The Smithsonian* magazine, offered a rough translation of the electronic code now speeding toward M13: "Here is how we count from one to ten. Here are five atoms that we think are interesting or important. . . . In some way this molecule [DNA] is important for the clumsy looking creature at the center of the message. . . . There are about four billion of these creatures on the third planet from our star. There are nine planets altogether, four big ones toward the outside and one little one at the extremity. This message is brought to you courtesy of a radio telescope 2,430 wavelengths or 1,004 feet in diameter." If anyone is listening, what impression do you think that message will give them regarding life on earth?

Like all respectable sciences, exobiology is more than just a series of experiments. The theoretical wing of the subject primarily concerns itself with two questions: What are the odds that ETI exists? What form and shape might this ETI take? Even though there are no definitive answers, the debate itself is illuminating. The mainstream of exobiologists begin by cataloging the number and age of stars in our galaxy, then factor in the possible development of planetary systems and the likelihood of life originating within them, and finally figure the probability of intelligent life evolving and the probable lifetimes of technological civilizations. When all this estimating is done, the best guess is that there are around one million technological civilizations in our galaxy alone! Even with the most conservative estimates of probability, the sheer number of star systems appears to make the existence of ETI extremely likely . . . somewhere. Since there are 250 billion stars in the Milky Way, this means that less than one star in 250,000 will have a planet inhabited by an advanced civilization. Without any clues to which stars are the best candidates, the search for ETI is on the same order as the search for the proverbial needle in a haystack.

Other, perhaps more daring exobiologists, think that these estimates are far too conservative. They start their inquiry from the second question: What will ETI forms be like? They point out that the mainstream only considers planets because their vision of ETI is carbon-centered, like life on earth, and planets seem the most likely habitat for this kind of life. This school rejects "carbon chauvinism" and proceeds to build science-fictional scenarios of life on suns, in the interstellar gases, and even within black holes. Given our current state of knowledge, they might just be correct.

One theoretician of the more imaginative kind, the eminent physicist

Freeman Dyson, has offered mind-stretching speculations concerning the most provocative possibility—supercivilizations far in advance of our own. Dyson and associate N. S. Kardashev have proposed a division of supercivilizations into three classes, on the logical basis of the evolution of energy-using technologies.

Earth is on the threshold of becoming a Class I society, one which consumes or produces an amount of energy of planetary proportions. When we extract energy from the weather, from earthquakes and volcanoes, from the spinning of the earth, and achieve controlled fusion, earth will reach the status of a Class I economy. The more advanced Class I supercivilizations would have the power to colonize solar systems, perhaps exchange information with other ETIs, and enter the lower echelons of Class II, where they would be able to exploit the energy of an entire star. The imaginative hypothetical artifact now known as a "Dyson Sphere" was offered as a hallmark of this class of development: The planets of a solar system could be dismantled and reassembled in a sphere around the central star, creating a 100 percent solar economy capable of fueling a culture millions of times larger and more diverse than our own.

A Class III civilization would be one capable of dealing with galactic levels of energy. Stars are used as tools; hospitable environments can be constructed at will; and the ETI would have the capability of ruling entire galaxies. Only the enormous energies of black holes, perhaps quasars, and other cosmic energy sources still unknown to us would make such a development possible.

The possibilities themselves offer a challenge to the limits of our imagination and our science. Only one thing is now certain in the science of exobiology: Either humans are the only life forms in the universe, or they are not. Whichever case proves to be correct, the proof will change us forever.

CONVERSATIONAL TACTICS:

What better topic for conversation than the topic of conversation with ETI? What do you say after "Take me to your leader"? One thing should be clear: Conversing in natural language is out. It is just too much to expect an inhabitant of Saturn's moons to speak English. That leaves mathematics or images. In 1972, the U.S. launched *Pioneer 10* through our solar system and out into deep space. It contained the following plaque. Assuming you were an ETI, how much information could you glean from this picture:

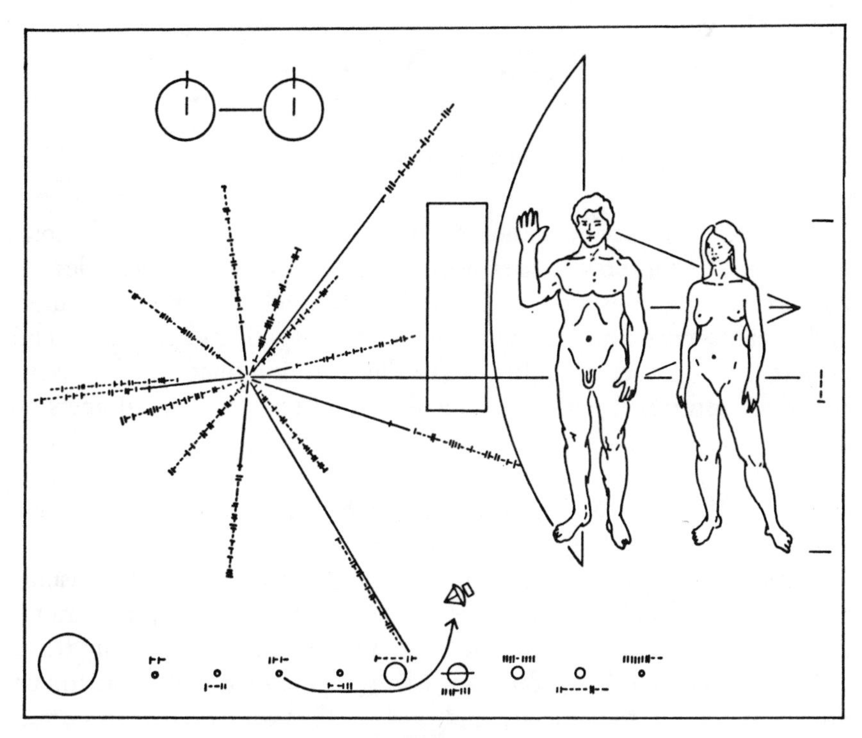

Exobiology—Earthlings' Message

RELATED TOPICS:

Bioethics, Black Holes, Consciousness, Image Enhancement, Origins of Life, Radio Astronomy, Viruses.

FURTHER READING:

Broca's Brain, Carl Sagan, Random House, 1979.

Intelligent Life in the Universe, Carl Sagan and I. S. Shklovskii, Delta, 1966.

The Quest for Extraterrestrial Life, ed. Donald Goldsmith, University Science Books, 1980.

FIBER OPTICS

DEFINITION:

A technology for communicating information through flexible, ultrathin, optical-quality glass fibers by using them as waveguides for modulated light. Fiber optics technology is combined with computers and lasers to create an integrated optics system with the ability to send and receive frequency-multiplexed, digitized information, coded by computer-controlled pulsed lasers and transmitted over optical fibers.

WHAT IT REALLY MEANS:

Optical fibers are nothing more complicated than tubes for transmitting flashes of light. What they really mean, however, are mind-boggling changes ahead. Communications revolutions, especially those which involve everybody, often start with simple inventions and end up changing life so dramatically and quickly that social structures crack under the strain. According to Alvin Toffler, the future shock phenomenon which now has us all dazed is directly related to the information overload forced upon us by our expanding communications system. Telephones, television, computers, satellites, and fiber optics are interlocking parts of the newly evolving solid-state consciousness grid.

Fiber optics will be nothing less than the nerve fibers of the quasi-artificial "world brain" we'll all know so intimately by the 1990s. One requirement for evolving a new kind of brain or building a better communication system is finding more efficient structures to handle information. Whether you are designing a telephone network or growing a nervous system, the ultimate goal is the capacity to transmit more and more complex messages at higher and higher speeds using less and less energy. The fiber optic revolution will multiply the present, already dizzying rate and density of message transmission several *millionfold!* Even more amazing in this inflationary age, the cost will *drop* as the capacity increases. Our lives will be changed in ways we can't even imagine now.

The best way to understand how a fiber optics communication network operates is to view it as a natural extension of the present telephone system. In the early telephone systems, a sensitive membrane in the headset translated the pressure changes that were caused by the human voice into a series of alternations in an electric current. The pattern of electrical changes passed along a series of copper wires and electronic switching stations to the receiver of the other telephone, where the electrical signal was converted back into the sound of a voice.

Nowadays, in order to accommodate all the telephone calls, computer-to-computer talk, and satellite transmissions, the mechanisms between the two telephone receivers have changed a great deal. Although voices are still converted from air-pressure fluctuations into electrical signals, the signals no longer travel as a smoothly varying flow of current; instead, they are turned into digital information—on and off pulses alternating at the rate of thousands per second—and then mixed up with hundreds of other conversations. Computers within the telephone system unsort all the mixed-together digital messages as they approach their destinations.

Even our sophisticated modern system has a limiting factor: the digitized flow of electrons over metal wires has a certain message-carrying capacity, and this restricts how thick and fast the messages can be sent. However, when pulses of *light* are substituted for the electrical charges, with optical fibers substituting for the metal wires, the upper limit for information flow is raised dramatically. The immense difference in information-carrying capacity (known as *bandwidth* in the jargon) is due to the much higher frequency of light waves. Audible sound is caused by air pressure oscillations in the 20–20,000 Hz range. Light, however, vibrates in the 10^{14} to 10^{15} Hz range.

At their higher frequencies, there is sufficient "room" in light signals to pile millions of separate frequency-multiplexed (chopped up, mixed together, and reassembled) streams of information. Your voice, along with the trillions of other communications which make the wired world go round, will soon be represented by a rapidly pulsing coherent light instead of the slowly fluctuating electrical current. A needle-thin laser beam and a hair's-width optical fiber are capable of transmitting entire libraries of information in less time than you take to read this sentence.

Certain engineering problems remain to be solved, but because glass is so much more abundant (and therefore less expensive) than copper, and because optical systems are lighter, more efficient, and more durable, you can expect the whole world to switch to fiber optics over the next two decades. If the printing press heralded the industrial revolution by putting scientific information in the hands of the public, what will be the effect of putting thousands of times more information at the disposal of billions of people?

CONVERSATIONAL TACTICS:

Fiber optics is the Swiss army knife of conversation—a portable, all-purpose tool that can meet almost any need. Discussing the movies? Many of the special effects created for science-fiction films are accomplished through fiber optics—that's how they're able to produce all the light points on those giant spacecraft. Discussing kitschy American culture? Those light trees which come in pleasing colors like chartreuse and mauve are products of fiber optics technology.

Talking about medicine? The small size, inertness, and flexibility of fiber optics allows less invasive diagnostic procedures. Patients can swallow a thin fiber in order to illuminate and photograph passages within the body, or surgeons can insert fibers through small incisions and then work them through the curves of the intestines or arteries. Remarkable films now exist, photographed from within such previously unexplored territory as the human circulatory system and the womb.

Fiber optics is most likely to be discussed as a communications medium. One of the earliest optical communications systems was employed in 1588, as the Spanish Armada moved up the coast of Britain. The British alerted neighboring towns by a series of bonfires on the headlands. This "system" transmitted only one, albeit crucial bit of information. In 1980, Atlanta Bell installed an optical system which transmitted 44.7 *million* bits of information per second. One can only hope that it will be as successful as the early British model. If you seek a patriotic counterbalance to the British innovation, don't forget that Paul Revere relied on a primitive digital optic system in the Old North Church: "One if by land, two if by sea . . ."

RELATED TOPICS:

Appropriate Technology, Digital Computer, Information, Lasers, Miniaturization.

FURTHER READING:

The Techno/Peasant Survival Manual, Colette Dowling, Bantam, 1980.
"Communicating on a Beam of Light," Lawrence Lessing, *Fortune,* March, 1973.

FISSION

DEFINITION:

The process by which certain atomic nuclei can be bombarded by neutrons and split, thus liberating the energy of the strong nuclear force, thereby yielding more neutrons and two atoms of lesser combined weight.

WHAT IT REALLY MEANS:

$E = mc^2$ (energy is equivalent to mass times the speed of light squared) is probably the most famous equation in science. Einstein's statement is as simple in principle as it is profound in implication. It means that what we think of as energy and what we perceive as matter are actually just two different states of the same entity. Furthermore, under the proper conditions, these two complementary states are convertible, one into the other. Because the c^2 term in the equation is a very large number (more than 34 billion), a very small amount of mass transforms into a great amount of energy. One of the means by which that tremendous energy can be unlocked is the process of nuclear fission. The other way of liberating the atomic force is nuclear fusion.

Particle physics paints a picture of the subatomic world which sometimes contradicts common sense. For one thing, the particles and aggregates of particles which make up matter are more illusory than our senses suspect. This piece of paper is mostly empty space, and the tiny entities buzzing around in that space are themselves slippery; the paper seems solid to us because the atoms and molecules move so fast that they blur into the illusion of solid matter. Another counterintuitive notion of nuclear physics is the idea that the "glue" which holds the particles together is just as real as the particles. The most potent glue is known as the *strong force,* and it is the energy of this nuclear cement which is liberated during nuclear fission—as much as two hundred million electron volts per atom!

The strong force only works within a range of 10^{-13} cm, roughly the diameter of an atom's nucleus, and it binds the nucleus together. Because protons have a positive charge, they always try to repel each other. The strong nuclear force in each atom must possess enough energy to keep all the protons and neutrons from flying apart. In the 1930s, scientists discovered that they were able to shoot free neutrons at the nuclei of especially complex atoms, such as uranium, and split them into two less complex atoms. The splitting became known as *fission* because it resembled the division of living cells.

Nuclear fission was experimentally observed before it was theoretically

Fission—The Liquid Drop Bomb

understood because, in the words of one physicist, splitting a nucleus with a neutron is about as likely as "splitting a brick with a peashooter." The uncontestable fact that this unlikely event does occur prompted a search for an explanation. The explanation nuclear physicists adopted was known as "the liquid drop model." The particles in the nucleus cling together the way molecules of water cling together in a drop. When a free neutron strikes the nucleus, an oscillation begins. Eventually this oscillation distends the nucleus into an elongated shape slightly larger than 10^{-13}cm. When this happens, the strong force is no longer powerful enough to hold the protons in the nucleus together, and their repulsive electromagnetic force takes over, distending the nucleus even further. Finally, the unstable atom becomes wasp-waisted and splits into two parts. The ability to initiate and control this fission process led to the atomic bomb and nuclear power.

Two concepts related to nuclear fission are often used in everyday conversation. The first is *critical mass.* During fission, free neutrons are knocked out of the nucleus and propelled into space. Three possible fates await these nuclear bullets: They can be absorbed into the nucleus of a stable atom, they may escape the fissionable mass, or they may strike another fissile atom and continue the process. Critical mass is the measure of the amount of fissionable material needed to ensure enough free neutrons to keep creating new fissions.

The second concept which has leaked from physics to social metaphor is *chain reaction*—a process related to the concentration of critical mass. Every time a neutron bullet splits an atom, two or three more neutrons fly. When enough new neutrons are propelled to keep the fission cycle going, a self-sustaining chain reaction occurs. A nuclear bomb is simply an ultrarapid buildup of neutrons, leading to a rapid and uncontrolled chain reaction. A nuclear power plant controls the chain reaction by using special materials as neutron absorbers. If this were to fail, a meltdown could occur.

CONVERSATIONAL TACTICS:

Nuclear weapons and nuclear power have been debated heatedly for decades. Will nuclear fission be the path to salvation or the road to destruction? We offer no further propaganda for either side of the question, but *Talking Tech* does have some advice regarding your Christmas shopping:

Fission and Superstition

(A cautionary verse for parents or children, appropriate to the Christmas season)

This is the tale of Frederick Wermyss,
Whose parents weren't on speaking terms.
So when Fred wrote to Santa Claus,
It was in duplicate because
One went to Dad and one to Mum—
Both asked for some Plutonium.
See the result: Father and Mother—
Without consulting one another—
Purchase two lumps of largish size,

Intending them as a surprise,
Which met in Frederick's stocking and
Laid level ten square miles of land.

MORAL:

Learn from this dismal tale of fission
Not to mix science with superstition.
—H.M.K.

RELATED TOPICS:

Energy Alternatives, Fusion, Meltdown, Particle Accelerator, Periodic Table, Relativity Theory, Toxic Chemicals.

FURTHER READING:

The Curve of Binding Energy, John McPhee, Farrar, Straus & Giroux, 1974.
"The Problems and Perils of Nuclear Energy," Kenneth F. Weaver, *National Geographic,* April, 1979.

FUSION

DEFINITION:

The process by which two light nuclei combine to form one new nucleus, of slightly smaller combined mass, with the missing mass converted into energy according to $E = mc^2$.

WHAT IT REALLY MEANS:

Nuclear fusion is the opposite of nuclear fission in several important respects. Instead of the strong nuclear force being liberated by splitting the nucleus, the atomic binding energy is released by forcing two atomic nuclei to combine; two nuclei are fused into one, and energy is released in the forms of heat and radiation. Fission requires extremely heavy, very rare isotopes of uranium and plutonium; fusion requires very light, relatively abundant isotopes of hydrogen and lithium. Furthermore, pound for pound, fusion can yield thousands of times more energy than fission. Because of these advantages, many people hope that the energy source which powers the stars will be harnessed for earthly consumption—before ten or twenty billion energy-eating humans greet the year 2000.

The problem with fusion is that it takes a tremendous amount of energy to get started. Stars, including our sun, are able to accomplish this because of their great mass; gravitational forces cause hydrogen nuclei to fuse into helium. On earth, we try to squeeze atoms by heating them. This means first having the technological ability to create a temperature of 100,000,000 degrees Kelvin and then the ability to contain that temperature long enough for the reaction to become self-sustaining. Thermonuclear fusion was first accomplished on earth by using a nuclear fission bomb as a detonator. Clearly, it isn't practical to use atomic bombs just to keep a fusion reactor going. Fortunately, more sedate methods are in the process of being developed.

At the moment, scientists are working on two different approaches to the problems of squeezing and confining fusion fuel. One approach uses magnetic devices. Hydrogen gas is heated until the electrons and nuclei separate; the resulting gas is a curious fourth state of matter known as *plasma.* In order to get the plasma nuclei to fuse, the fuel must be confined long enough for the reaction to catch on—something like the critical mass in fission reactions. The big engineering problem is that no known substance can hold plasma at temperatures of 100,000,000 degrees Kelvin, so one branch of fusion research uses powerful magnetic fields to create a nonmaterial container for the plasma, a "magnetic bottle." These doughnut-shaped containers are known as *Tokamaks.*

The second approach to fusion is known as *inertial confinement.* Instead of holding the fuel in place by confining plasma for a relatively long time, hydrogen isotopes are sealed in tiny glass beads, then zapped with laser energy from all sides. The most powerful multi-beam laser now used is named *Shiva,* after the Hindu goddess of creation and destruction, and can only operate at trillionth-second intervals, because such a tremendous amount of energy is needed to power that zap. Theoretically, Shiva can put out more than twenty-five *trillion* watts of power—over fifty times the total output of every power plant in the U.S.! Because of these technological difficulties, the first operating nuclear fusion plant—using either magnetic or inertial confinement—is not expected until the 1990s, at the earliest.

CONVERSATIONAL TACTICS:

Question: What is the connection between a cup of seawater, the interior of a star, and the price of petroleum?

Answer: Nuclear fusion, of course.

If fusion is good enough to fire up the cosmos, it's good enough for our energy needs, say proponents of fusion-power research. Consider the matter of fuel. The uranium and plutonium needed for fission reactors are notoriously rare and expensive substances, to say nothing of their toxic effects. The preferred fuel for fusion is deuterium, which is found in seawater. This means that one gallon of seawater could provide as much energy as three hundred gallons of gasoline. A water main only twenty inches in diameter could carry the fuel for the entire planet, if we only knew how to separate the deuterium and force it to fuse.

Fusion research is serious business, but at least one respectable authority has used humor in his presentation. Physicist Robert Bussard has been exploring the concept of "throwaway fusion reactors." The idea is to bypass the problems of designing a reactor to hold up for years under fusion conditions by creating magnetic bottles strong enough to last a month or two; when they wear out, you throw them away and plug in new ones! The idea is serious, but Bussard's original presentations for the low-budget Tokamak were made with a model built out of a *bagel.* You can reproduce one of the most exciting new ideas in fusion research by purchasing a bagel at your local delicatessen, toasting it, then coiling wire or yarn through the hole in the center. The bagel is roughly the shape of the confined plasma, and the wire represents the magnetic bottle. Just such a model, painted green, once actually sat on the desk of the man in charge of the nation's magnetic-confinement fusion research program.

RELATED TOPICS:

Energy Alternatives, Fission, Lasers, Neutrinos, Relativity Theory, Stellar Death, Sunspots.

FURTHER READING:

"Disposable Fusion Reactors," Janet Raloff, *Science News,* August 12, 1978.

"Fusion Power's Utopian Promise," Karen Ray, *Science Digest Special,* Winter, 1979.

GALAXY

DEFINITION:

A large-scale aggregate of stars, gas, and dust, having either a globular, spiral, elliptical, or irregular structure and containing an average of one hundred billion solar masses.

WHAT IT REALLY MEANS:

If you live in a rural area, away from smog and city lights, you can partake of a visual experience that awed our ancestors. Look at the sky on a moonless night and note one broad band of stars so thickly clustered that it resembles a milky spill from horizon to horizon. The Greeks also perceived this as a belt of milk in the sky and called it *galaxias,* after their word for milk; we now use the word *galaxy* to denote a star system. The Romans called that grand corridor of light the *via lactea,* or road of milk; we now call our home galaxy the Milky Way. What those ancient astronomers couldn't know was that some of those tiny, unfailing beacons were themselves distant groups of stars.

In the mid-eighteenth century, a British instrument maker, Thomas Wright, suggested that the Milky Way was actually a collection of stars composing a vast disc in space. He proposed that our terrestrial viewpoint is from within part of the disc, so that it is natural for us to observe a thick star cluster along the plane of the vast "grindstone." It turns out that Wright was right. The model of our galaxy now accepted by astronomers is that of a fairly flat disc of stars with a globular mass in the center. The disc has a diameter of eighty thousand light years, a thickness of six thousand light years, and a spherical halo of one hundred thousand light years' diameter. Our own solar system is located in the stellar suburbs, around thirty thousand light years from the center.

It was obvious, even to the ancients, that we live in an archipelago of stars. One can't dismiss a map of the universe as large as the Milky Way. What didn't occur to the first stargazers was the possibility that some of those light points were themselves star clusters, some of them much larger than our own. The *nebulae* ("little clouds") which triggered such outlandish speculations were named by a tenth-century Persian astronomer, Abdurrahman Al-Sufi. Eight hundred years later, the German philosopher Immanuel Kant merged Wright's theory about our home galaxy with Al-Sufi's discovery of fuzzy objects that weren't stars. Kant suggested that some of the nebulae might be galaxies like our own.

It took another couple of centuries before the argument was finally

Galaxy—Where Are We?

settled observationally. In 1923, Edwin Hubble used the new one-hundred-inch telescope on Mount Wilson to resolve the nebula M31 into separate stars. One effect of modern astronomical techniques is that the size of the universe keeps growing as our observational tools improve. We now know that the Milky Way is only one of a hundred billion galaxies and that all the galaxies are rushing away from each other, some of them at speeds close to 90 percent the speed of light.

Discoveries in nuclear physics and cosmology have been added to the observations furnished by astronomers in an attempt to create a probable scenario for the original creation of stars and galaxies. According to the big bang model of creation, the primordial singularity exploded with such force fifteen billion years ago that the hot cosmos took one billion years to cool enough to gestate star systems. In the beginning, all the matter in the universe was in the form of a superheated, expanding cloud of hydrogen. When the primordial gases cooled enough to condense, gravity and pressure fluctuations within the vast clouds began to draw the gas particles close together into giant protogalaxies.

After millions of years of contraction, the gas heated up again through the friction caused by this intense gravitational compression. At a critical temperature near 20,000,000 degrees Fahrenheit, hydrogen began to form helium, through the process of thermonuclear fusion. The implosion caused by gravity was balanced, in certain centers of condensation, by the explosion caused by fusion—millions of H-bombs' worth of explosion. The entities we know as stars were a result of that implosion-explosion balancing act.

CONVERSATIONAL TACTICS:

Galaxies are such enormous objects that it often helps to bring them down to a human scale. While you explain the milky origins of the word *galaxy*, you can model the forces which formed the galaxies into their present shapes by stirring a cup of hot coffee and adding a dollop of chilled milk. In those S-curved eddies, double-spiral whirlpools, and irregular globular clusters—the proverbial "clouds in your coffee"—are reflected the unthinkably larger forces of convection and turbulence which formed the great spinning discs of stars. The coffee represents the gravitational vortices forming the protogalaxies; the milk symbolizes the vast clouds of gas whence we came.

RELATED TOPICS:

Black Holes, Cosmology, Fusion, Quasar, Radio Astronomy, Red Shift, Stellar Death, Sunspots.

FURTHER READING:

Violent Universe: An Eyewitness Account of the New Astronomy, Nigel Calder, Viking, 1969.

Galaxies and Quasars, William J. Kaufmann III, W. H. Freeman, 1979.

GÖDEL'S INCOMPLETENESS THEOREM

DEFINITION:

A basic result of mathematical logic which demonstrates that any logical system complex enough to do arithmetic must contain a fundamentally undecidable proposition.

WHAT IT REALLY MEANS:

There is a natural tendency to assume that modern science is forever widening our scope of knowledge—we see farther into the heavens, deeper into the atom, and understand more about ourselves than ever before. Yet three of the twentieth-century's greatest discoveries—the second law of thermodynamics (entropy), Heisenberg's uncertainty principle, and Gödel's incompleteness theorem—are considered great precisely because they all put definite limits on what we may know.

These three theoretical limits are all "impossibility results," each asserting in its own way that no matter what conditions prevail, certain things can never be known. Contrast this with the typical scientific statement: Salt is composed of an acid and a base; black holes are collapsed stars; the current position of land masses is the result of continental drift. All of these statements are empirical and may be confirmed, or ruled out, by experimentation or observation. But an impossibility claim, because it refers to all possible preexistent conditions, cannot be empirical. It must be logical, the result of a pure intellectual exercise. Gödel and Heisenberg told us as much about how our minds work as they did about the workings of mathematics, matter, and energy.

Like most scientific results, Gödel's theorem did not spring up without any historical antecedents. In fact, Gödel's 1931 paper ("On Formally Undecidable Propositions of Principia Mathematica and Related Systems") ended an almost two-thousand-year-old mathematical dream. Euclid may be memorable to us because of high school geometry, but to the professional mathematician he is more important because of his method of *axiomatization.* (You start with a few "obviously true" statements, the *axioms,* and manipulate them, using a few "self-evident" rules of inference, in order to prove new *theorems.*) Euclid set down the method of proof which mathematicians follow to this day.

Using Euclid's geometry as a model, centuries of mathematicians labored with two goals in mind: first, to reduce all mathematics to a single, axiomatized system; second, to demonstrate that this system was "correct" and that embarrassments like non-Euclidean geometry would not reoccur. By the end of the nineteenth century, the first goal had been met, and the hottest mathematical questions dealt with the relationship between logic and mathematics. The general hope was that all mathematics could be derived from an axiomatized logical system, thus fulfilling the Euclidean hope. Gödel demonstrated that it was a vain hope; he proved that *no* axiomatized logical system for mathematics could ever be "correct."

If one is going to go to all the trouble of axiomatizing a system, there are a few criteria which any system claiming to be "correct" should meet. First, it should be *sound.* You don't want the system to prove any false theorems. Second, it should be *complete.* You want the system to prove every true theorem. Third, it should be *decidable.* For any arbitrary formula, you want the system to tell you if it is a theorem or not. Gödel showed that any axiomatic system which contains the elements to do arithmetic must be undecidable because there will always be two formulae, G and −G (not G), which the system cannot show to be theorems. Furthermore, since either G or −G must be true, the system must also be incomplete. In other words, certain mathematical truths lie outside the realm of specific axiomatizations.

Although Gödel's original paper was long and complicated, as with many great insights the logic is simple. Gödel's formula "G" simply represented a mathematical way of saying: "I am not provable." If proven, G would be false and the system unsound. If unproven, G would be true and the system incomplete. Either way, the system is bound to be not "correct." By importing the paradoxical quality of ordinary language into an abstract, precise mathematical system, Gödel demonstrated that mathematics and the mind cannot be separated.

CONVERSATIONAL TACTICS:

The key to Gödel's proof is the property of *self-reference;* "G" is talking about itself. When we turn our language back on itself, strange things happen. Consider the claim, "I am lying." If it's true, then it must be false, since I can't be both telling the truth and lying. If it's false, then it must be true, since if I'm not lying I'm telling the truth. Such paradoxes are known as *antinomies,* and they create a serious problem for logicians and philosophers. Seeming paradoxes also form a large set of thought puzzles. If you think you understand the "liar paradox," try out "the anthropologist's dilemma."

An anthropologist is studying an island on which two tribes reside. He knows that one tribe always tells the truth, while the other one always lies. He also knows that the truth tellers all live in the north while the liars all live in the south. Unfortunately, during the day the tribes work all over the island, and the anthropologist is never sure if he is talking to a northerner or a southerner. One day, the anthropologist must find a truth teller to transmit important information to an assistant. What logical procedure could the anthropologist employ to insure that any information would be transmitted correctly?

All the anthropologist need do is ask one native to ask another, "Which side of the island do you live on?" If the questioner tells the anthropologist, "He says he lives on the northern side," then the questioner must be a truth teller, because if the responding native had lived in the north he would have said so. If he had lived in the south, he would have lied and said he lived in the north. Either way, the questioner—if he was a truth teller—would say north.

RELATED TOPICS:

Artificial Intelligence, Digital Computer, Entropy, Infinity, Mathematical Modeling, Uncertainty Principle, Zeno's Paradox.

FURTHER READING:

Gödel, Escher, Bach: An Eternal Golden Braid, Douglas R. Hofstadter, Basic Books, 1979.

The Mathematical Way of Thinking, ed. James R. Newman, Simon and Schuster, 1956.

HOLOGRAPHY

DEFINITION:

A technique which uses wave-front reconstruction to produce apparent three-dimensional images.

WHAT IT REALLY MEANS:

Holography is one of those prodigious technologies that occasionally sprout up unpredictably out of basic research and grow to alter our lives in unforeseeable ways. In 1947, a physicist named Dennis Gabor devised a technique for improving the images captured by the newly invented electron microscope. Gabor dubbed his discovery *holography,* after the Greek words for "whole writing," meaning that these images reproduce the entire message and not just one part. For sixteen years, holography was discussed only in technical journals. Then, in 1963, two electrical engineers combined the Gabor technique with an even younger invention—the laser—and a potent new way of seeing was born.

Holography remained obscure for so many years because it had no immediate application to daily life for most people. It started out as a way to create images from electron beams—an exciting prospect to molecular biologists but not half as much fun as color television to the nonscientist. The idea of applying the same technique to visible light instead of electrons was simply not a thinkable concept in 1947—no light which existed in nature or the laboratory could meet the technical requirement for the process. The application of an entirely new kind of light instantly raised the concept from obscurity. Emmet N. Leith and Juris Upatnieks made front-page news around the world when they created an apparently three-dimensional image with the hybrid technology of laser holography. Their first image, Leith later recalled, was nothing more dramatic than a "six-inch pile of junk we had lying around the lab."

The images, if not the scientific elegance of the method, immediately captured the public imagination. When true 3-D monster movies didn't open at the corner theater, public expectations were deflated. Because of these premature forecasts, many people now think of holography as one of the failures of the new technology, but the true impact of the process lies in its information-encoding power, and in this respect the most important social effects of the holographic revolution have yet to hit us.

The unique capabilities of holography, and the physical principles by which it works, can best be understood by comparing it to photography. In photography, if you want to reproduce the image of an apple, the apple is first illuminated by a flashgun or natural light; a portion of the light is reflected off the apple and toward the camera, where the lens focuses an image on a photosensitive emulsion. The way the apple reflects, absorbs, and scatters light creates variations of color and brightness which cameras (and our internal optical systems) translate into an image. The lens "freezes" a cross section of that reflected light into the focal plane.

If you were to take a holograph instead of a photograph, you would shoot a beam of laser light at the apple, but before it hit the apple, the beam would hit a half-silvered mirror. Half the beam would go through the mirror to reflect from the apple back to the camera, just as in photography. But in holography that apple-reflected beam intersects with the other beam reflected from the mirror—the *reference beam*—which does not bounce off the apple. The intersecting beams are then directed through the aperture of a lensless camera. The intersection of those two beams is the secret of holography, for it takes advantage of the physical properties of light to capture information from many more angles than a single focal plane.

Because of the wavelike aspect of light, two intersecting beams cause *interference patterns.* Like ocean waves, light waves reinforce each other when peaks and troughs of the waveforms are in alignment and neutralize each other when they are out of alignment. In a hologram, a third pattern is created from the two intersecting laser-wave patterns: where the waveforms are in synchrony, more light will be reflected from the image, and where they are out of synchrony, less light will be reflected. A hologram film does not record the light variations in a single plane, as in a photograph. Rather, it records bands of light and darkness. These black and white bands correspond to the interference pattern produced by the light reflecting from the apple in combination with the reference beam. All the information needed to make a photographic image, and more, is encoded in those light and dark bands.

The reference beam is the key to reconstructing a holographic image from an interference pattern. It's like solving an algebraic equation. If you know the waveform characteristics of the reference beam, and you have the interference pattern, it is possible to solve the equation, crack the code of the light and dark bands, and reconstruct an image of the apple—an extraordinary kind of image, in fact. If the same kind of light beam which was used to make the hologram is shined through the interference pattern on the film, the colliding light-wave patterns "unsort" themselves into an image.

The reason this visual reconstruction can work so precisely is that the reference beam is a very special kind of light. Laser light, because of the way it is produced, contains only light waves of a single frequency, all in synchrony with each other; this *coherent* light is unvarying enough to act as a constant in reconstructing the image. If light is pure information, as McLuhan claimed, then laser light is information *amplified.*

If you were to develop and compare your photo and holo of an apple, several curious differences would emerge. Look at the photograph. It shows you a single plane of the apple; you can never see the sides or back of the photograph. In comparison, if you peek from one side or from the top of your hologram, you would see angles not represented in the straightforward view. Because there is no lens in a holographic camera, the point of view is not frozen into a single focal plane. Those interference patterns on the hologram contain more visual information about the apple than could be contained in a single plane of ordinary reflected light.

When you try to mutilate a hologram, its properties reveal themselves to be as amazing as they are amusing. Cut your photograph in half, and you have half an image of an apple. Cut your hologram in half, and you will see an image (slightly fuzzier) of a whole apple. Cut the hologram into a dozen pieces, shine a reference beam through one of the fragments, and you will get an even more blurred but still intact image. You can scratch or pierce the surface and still see an ummarred image. Dust or stains on the film surface won't affect the projected picture. At this point, the true potential of holography begins to suggest itself: Wave-front reconstruction is a highly reliable way of storing a very large amount of information in a very small amount of space and retrieving it quickly.

The designers of the next generation of digital computers, and their colleagues in the artificial intelligence field, are seriously considering the advantages of holographic information storage, especially now that integrated optic systems are being designed to use light waves instead of electronic pulses as memory elements. At the most ambitious end of the speculative spectrum, holographic memory could make it possible to build a portable thinking machine. At the more mundane level, it would be possible for an automobile repairperson to carry a twenty-volume service manual in a pocket-sized holo-microfiche reader.

CONVERSATIONAL TACTICS:

The most common image of the hologram is of a three-dimensional picture. This technological artifact, fascinating though it may be, is perhaps not the most important feature of holography. The conversational aspect of holography may well become as important to its development as the creation of new kinds of hardware, for a hologram is also a powerful metaphor, a useful model for some of the most complex things on earth. There is at least one other place in nature where the entirety of a message is contained within each of its component parts: Every living cell contains within its DNA code the blueprint for constructing the entire organism. Another vitally important but still mysterious phenomenon that seems to manifest itself in a holographic manner is human memory.

Psychological theories about human memory have long been divided between two points of view. Is memory a *localized* function: Does a day in May exist in a specific group of brain cells? Or is memory a *general* function: Is a day in May filed in many places, under smells, sights, sounds, and associations? Experimental evidence has been accumulated in support of both points of view. One of the most talk-provoking new theories is that of Karl Pribram, distinguished Stanford psychologist, who proposes that the processes of memory and other higher mental functions are varieties of holographic reconstruction accomplished by our nervous systems.

When the sight of a flower in wintertime evokes the detailed recall of a spring day years ago, your memory may be functioning holographically. In Pribram's theory, the first stimulus (the sight of a flower) sets off a resonance through the brain's store of holographically coded memories (a day in May, don't forget to buy flour at the store). The proper parts of particular memories are quickly reconstructed by your brain's own "reference beam." Some psychophysiologists think the "alpha rhythm" detected by electroencephalographs may serve this purpose.

RELATED TOPICS:

Consciousness, Fiber Optics, Image Enhancement, Information, Lasers.

FURTHER READING:

Toward a Science of Consciousness, Kenneth R. Pelletier, Delta, 1978.
Languages of the Brain: Experimental Paradoxes and Principles in Neuropsychology, Karl H. Pribram, Prentice-Hall, 1971.
"I Can't Believe I Saw the Whole Thing," Isaac Asimov, *Saturday Review of Science,* September 2, 1972.

HOT-BLOODED DINOSAURS

DEFINITION:

A recent hypothesis that some dinosaurs were endothermic—capable of internally generating their own body heat.

WHAT IT REALLY MEANS:

Dinosaurs! Why is the human imagination so captivated by an extinct species, of all things? The reconstructed fossil remains of these gigantic beasts are the main attractions in museums of natural history; dinosaur skeletons are among the best-selling model kits; and thousands of people pay their own way each year for the dubious privilege of digging for bone fragments in the hot sun of desert wastelands. Carl Sagan has even suggested that children's nightmares and fears of "monsters" are evolutionary vestiges of our ancestors' adaptive responses to predatory dinosaurs. Who were these behemoths that we now find so fascinating? What were they like, and why are they gone?

Scientific speculations regarding these primordial giants began 160 years ago with the discovery of the first giant reptile fossil find. Twenty years later, in 1842, the original creatures that had left the fossils were dubbed "dinosaurs," from the Latin roots for "monstrous lizard." As more fossils accumulated, and as better techniques for dating and reconstructing them were developed, a standard picture of the dinosaur epoch emerged: They lived during the Mesozoic era, which began 225 million years ago and ended 65 million years ago. The earliest dinosaurs were descended from smaller, two-legged reptiles known as thecodonts. Given 160 million years to evolve, a great variety of dinosaur forms flourished: *Brontosaurus,* the huge, thirty-ton herbivore which had to return to all four legs and was literally a "pea brain"; *Tyrannosaurus,* the forty-foot-tall killing machine with the four-foot jaw (also known as "the one that looked like Godzilla"); *pterodactyls,* which glided on batlike wings; and hosts of armored, spiked, bony-crested, swimming, creeping, crawling, and stomping versions.

Even with all that diversity, the common wisdom has long held that all dinosaurs shared similar reptilian characteristics: the use of a hard-shelled egg to contain the embryo; a scaly, leathery hide; and a cold-blooded or ectothermic metabolism in which external heat sources are needed to raise internal temperature.

Within the last decade, this standard version has been questioned as new fossils were discovered which did not fit neatly into our old conception of the dinosaur as a slow-moving, cold-blooded reptile. The *deinonychus,* or "terrible claw," was a relatively small dinosaur standing about five feet tall and weighing around a hundred and fifty pounds. Everything we now know about it—its anatomy (which enabled it to run quickly), good vision, a dynamic tail for balancing, and pivoting hind talons for slashing—indicates that it may have been a warm-blooded predator. Another fossil, *Archaeopteryx,* the first bird, also supports the warm-blooded dinosaur hypothesis. Birds are warm-blooded, and if birds were descended from dinosaurs, there may well have been other warm-blooded dinosaurs.

Further evidence is provided by microscopic study of dinosaur bone tissue, which is more mammalian than reptilian. Finally, population estimates may indicate a high predator-to-prey ratio. If this turns out to be the case, this would show a high rate of food consumption among the predators—another warm-blooded characteristic. If it is conclusively determined that some dinosaurs were endothermic, then many of our other notions regarding them—and all of evolution—may have to be rethought. Brain size, social behavior, and adaptability in our own species is thought to have developed rather late in our own evolution; if our ideas about earlier species are incorrect, theories about human evolution may be affected.

Of course, 140 years of what the famous biologist C. H. Waddington calls COWDUNG—the Common Wisdom of the Dominant Group—will not yield so easily to a few fossil remains. Anyone claiming to prove the endothermic hypothesis will have to provide answers to two challenges. First, how could a warm-blooded dinosaur consume enough food to maintain its body temperature? A human consumes about forty times as much food as a human-sized lizard, and 80 to 90 percent of the energy released is needed to maintain a constant temperature. It seems

Hot-Blooded Dinosaurs—Reptile Love

improbable that a brontosaurus could eat and digest food quickly enough. A second, more intriguing question is: Why didn't any dinosaurs survive the Mesozoic era? It seems unlikely that *all* warm-blooded dinosaurs would have failed to adapt to changing environmental conditions, since one of the advantages of endothermic metabolism is greater adaptability.

CONVERSATIONAL TACTICS:

Whether or not they were warm-blooded, the dinosaurs seem to have disappeared from the evolutionary stage with undue haste. In at least one way which relates directly to the survival of our own species, the reasons for the dinosaur's sudden demise are of interest today. How did a species which dominated the earth for over one hundred million years (our own Darwinian twig is only a paltry few million years old) suddenly disappear sixty-five million years ago? Could it happen to us? Should you find yourself involved in a discussion of "dinosaur death," here is a list of current theories which purport to explain what happened to them:

1. *Constipation theory:* A change in the plant life at the end of the Mesozoic era eliminated a natural laxative from the herbivores' diet. Once they all died off, there was nothing left for the carnivores, and they too perished.

2. *Poison plant theory:* About 120 million years ago, the first flowering plants appeared. The plants produced alkaloid toxins such as strychnine which poisoned the herbivores, again leaving nothing for the carnivores to eat.

3. *The egg-depletion theory:* The Age of Mammals succeeded the Age of Reptiles; this theory proposes that as mammals became more numerous they ate increasing amounts of dinosaur eggs, which leads to the question: Which died first, the dinosaurs or their eggs?

The above three possibilities may have been contributing factors, but they do not seem catastrophic enough, nor quick enough, to have ended the reign of the dinosaurs so abruptly. Variants of the following three theories seem to have greater explanatory power:

4. *Weather theory:* We know that the earth's climate is constantly shifting between warmer and cooler periods. This long-held theory asserts that the climate became significantly colder about sixty-five million years ago; the dinosaurs couldn't adapt. Obviously, if the warm-blooded hypothesis is correct, this theory is probably incorrect.

5. *Extraterrestrial radiation theory:* Recent fossil findings indicate that dinosaur eggs were becoming increasingly thin-shelled. This theory proposes that a supernova bathed the earth in high-energy radiation, killing many dinosaurs outright and rendering the rest incapable of having offspring.

6. *Monstrous meteor theory:* Nobel physicist Luis Alvarez proposed what is currently the most scientifically popular theory—that a giant asteroid collided with the earth, throwing so much dust into the atmosphere that it blocked the sun for years, destroying plant life and starving the larger life forms. Alvarez based his hypothesis upon the finding that iridium, a metal uncommon on earth but abundant in

extraterrestrial objects, is found in relatively large amounts in the geological strata which coincide with the time of "the great dying."

Although it sounds like a grade-B science-fiction movie, the powers that be are taking this theory seriously. In February, 1981, NASA's advisory council proposed keeping track of the eight hundred known larger asteroids and preparing a nuclear-armed space missile to divert such objects before another such collision puts humans on the list of extinct species. The NASA panel suggested that the program, named Operation Spacewatch, might also help avert accidental nuclear warfare by tracking improbable sources of blips on defense radar.

RELATED TOPICS:

Human Origins, Origins of Life, Photosynthesis.

FURTHER READING:

"New Ideas About Dinosaurs," *Nationa Geographic,* August, 1978.
The Hot-Blooded Dinosaurs, Adrian J. Desmond, Dial Press, 1975.

HUMAN ORIGINS

DEFINITION:

The application of the theory of evolution to the origin of *Homo sapiens.*

WHAT IT REALLY MEANS:

The most disturbing and powerful question science can attempt to answer is a simple one: "How did the human race originate?" The scientific inquiry into human origins involves us all in a supreme paradox: We pride ourselves on our consciousness, which raises us above the beasts, but our science, which is the primary evidence for the superiority of our intelligence, has told us that our ancestors were apelike creatures. The general tenets of Darwin's theory of evolution were the cause for violent debate a century ago, but modern scientists accept Darwin's principles and are now attempting to refine the picture of our evolutionary heritage.

If we are indeed the descendants of furry forest beasts, how exactly did our particular branch of evolution happen to produce featherless bipeds like us? The answer to that question, although rooted in the distant past, involves issues which are vital to us all today: Are we the heirs to an evolutionary legacy of violent aggression or the descendants of creatures who were skilled in thinking and cooperating? If we can discover who our forebears were, and what they had to do to survive, we may learn something crucial about how four billion humans can survive in the near future.

Use the most powerful observational tool at your disposal—your imagination—to view our long-term history from a comfortable perspective. Instruct your mind's eye to zoom back from the twentieth century to the primordial past, and start visualizing the first frame of this mental time-lapse scenario at a relatively late point in biological evolution, the end of the era of the dinosaurs.

About seventy million years ago, the reptiles rapidly faded from their long period of dominion. Our earliest ancestors, the mammals, scurried from their hiding places to occupy the niches vacated by the extinct reptiles. Some returned to the sea and became whales or dolphins, others took wing and became bats, some stay-at-homes remained on the ground . . . and one group took to the trees. Zoom your mental camera lens closer to the tree-dwelling mammals, and note their binocular vision (for judging the distance from one branch to another), the hands for grasping branches, and the extra brain tissue devoted to coordination, prediction, and balance. It is here, in the trees, that the story of the human race truly began.

About fifteen million years ago, the earth's climate began to change, growing colder and drier over wide areas. Dense forests receded, vegetation changed, and open woodlands surrounded by grassy savanna became prominent ecological features. One survival strategy adopted by certain species was to leave the safety of the trees to strike out on their own in the new grasslands. *Ramapithecus* ("Rama" from a mythological Hindu hero, and "pithecus" from the Greek word for ape) did just that, and became the original *hominid,* the first member of the human family.

Although only three feet tall, weighing just forty pounds, and possessing a relatively small brain, *Ramapithecus* had a number of characteristics which made him our most likely relative: His dentition, particularly the reduced size of his canines, was remarkably humanlike, and his brain, while only about 250 cubic centimeters, began to have recognizable frontal lobes. Scientists conjecture that survival in the open required a sustained, erect posture, so *Ramapithecus* may have taken the first step that would end with us, the only truly bipedal mammals.

Unfortunately for paleontologists, our little ancestors became extinct about eight million years ago. It was not until approximately five million years ago that our next relative, *Australopithecus* ("southern ape"), made his appearance. This gap in the fossil record has led to the popular notion of a missing link—a hypothetical animal assumed to connect apes to humans. The existence of such a creature is crucial to those who subscribe to what Stephen Jay Gould of Harvard has called "the ladder metaphor" of evolutionary change—evolution as a continuous sequence of ancestors and descendants. However, as Gould points out, evolution does not proceed linearly like a ladder; rather, its structure is one of *branching*, more like a tree or bush.

Speciation—the splitting off of a new lineage from a parent stock—is the mechanism of evolutionary change. Whole species don't necessarily evolve; major changes may occur in small, isolated populations where genetic change can spread rapidly. We are less like the top rung of a ladder that began with *Ramapithecus* and more like the top, healthy branch of a tree with *Ramapithecus* at the roots.

With the advent of *Australopithecus*, the bush sprouted a few new branches. We do not know exactly how many types of *Australopithecus* may have coexisted, but two, A. *robustus* and A. *africanus*, have left the best fossil record. *Robustus* was nearly our equal in size, standing almost five feet tall and weighing as much as 130 pounds. His brain was almost twice the size of *Ramapithecus*'s—500 cc. The most interesting fact about *robustus* is the evolutionary stability of his brain size. Although we have unearthed fossils spanning a million years, his brain never changed in size. In this sense, he represents an evolutionary dead branch.

Africanus stood only three to four feet tall and weighed between forty-five and sixty-five pounds. He may have predated *robustus* by two million years and lived long enough to compete with him. The secret to his success lay in the development of his brain over successive generations. During his period of development, his brain apparently increased in size, the largest A. *africanus* achieving about 600 cc of precious gray matter. Studies of *endocasts*—molds of cranial interiors—show increased growth in association areas known to be related to higher-level analysis and language. *Africanus* was also the first hominid to use stone and bone tools.

By the end of the age of the Australopithecines, the brain-tissue expansion was well into its most explosive period of growth. The exact fate of the Australopithecines is not known, but it appears that the robust and hyper-robust branches withered, leaving the future to the smaller, brighter cousins. Approximately three and a half million years ago, *Homo habilis,* the first of our genus, appeared. About midway in size between the two Australopithecines—four and a half feet tall, sixty-five to one hundred and ten pounds—their brains were nearly one third larger, weighing 800 cc. Their skulls also indicate higher foreheads, which suggests a significant development of the neocortical areas in the frontal and temporal lobes—a truly human characteristic.

Habilis used tools with enough regularity for anthropologists to piece together a probable scenario of *Homo habilis* life that exhibits even more humanlike aspects: widespread tool use, division of labor, food-sharing, and home bases rather than constant foraging. To scientists, *habilis* is a distant cousin. You probably would prefer to see one in a zoo than at your dinner table, however.

About one and a half million years ago, the first of our recognizably human relatives appeared. *Homo erectus* was nearly our size—one hundred to one hundred and eighty pounds, four and a half to six feet tall—and had a high forehead combined with a smaller jaw. The facial configuration took on what an anthropocentric bias would call "a less brutish cast." Most importantly, his brain size of 1250 cc overlapped with ours. In less than one million years, nearly a pound of brain tissue evolved—an astonishing rate of growth, considering the billions of years which had gone before. The environmental pruning process began to switch from selecting those individuals with larger brains to selecting those who learned what to do with their cranial equipment.

Homo erectus might also be called *Homo technologicus.* Not content with randomly chipped tools, he produced tools in specific shapes, tools with double edges, and the first hand axes. Some anthropologists believe that making tools of a regular design may have been only one of a number of related behavior patterns which include food taboos, kinship customs, and certain rules of language. It is known that *erectus* was the first hominid capable of altering the local environment through the use

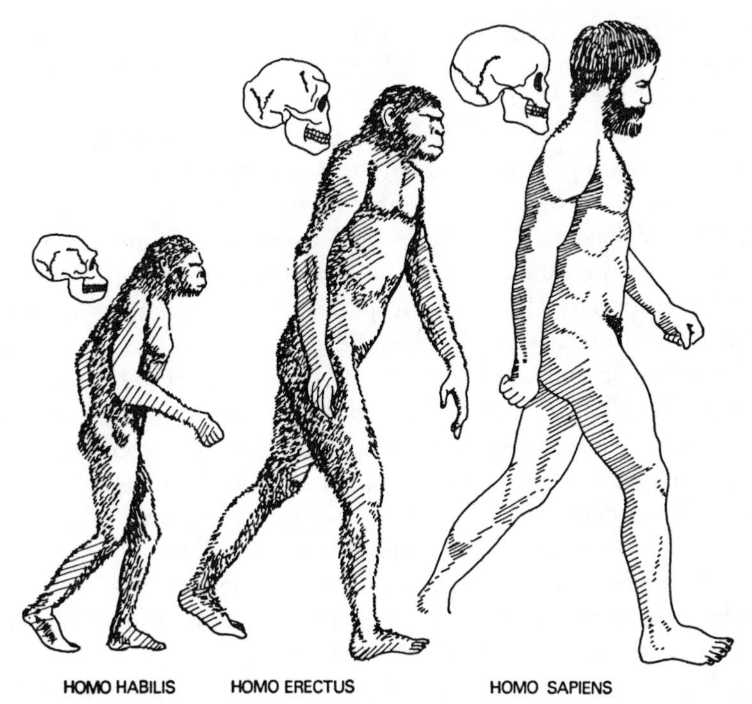

Human Origins—The Human Family

of fire, clothing, and shelter. When those big-brained toolmakers began to manipulate nature and teach each other the vital tricks of the survival trade, culture, rather than environmental forces, became our species-shaping tool.

The rate of *Homo erectus*'s brain growth peaked about half a million years ago and lasted altogether less than a quarter of a million years. During that extremely recent evolutionary epoch, *Homo sapiens* appeared. With brains as large as 2200 cc, we have the physiological capacity to live up to our name—"wise man." If we do, perhaps this story will have a happy ending. If we succeed in surviving the next few dangerous generations, *Homo sapiens* and successors might even rule as long as the dinosaurs did.

CONVERSATIONAL TACTICS:

What triggered the sudden expansion in brain size? What environmental pressures made our ancestors' brains grow those crucial extra layers of cells in their skulls? Why and how did consciousness and civilization emerge when they did? There are so many unknown areas in that critical period that there is still room for a range of hypotheses. Conversationally, the hypothesis of human origins which specific people favor is almost a personal litmus test of who they think they are.

There are currently three competing scenarios—the *tool-using, hunting,* and *food-sharing* hypotheses. All three schools are Darwinian in agreeing that the human mind, just like thumbs and teeth, is a product of the evolutionary history of our species. But what human invention or habit forced us to *think* so hard, so fast?

The oldest school asserts that those of our ancestors who knew how to make more effective tools tended to survive and procreate at a greater rate than their competitors; more important, the tool users taught successive generations how to make better tools. Language and learning were the ultimate tools, for they overturned natural selection—the invention of culture was an evolutionary force. The information-using skills associated with language and toolmaking were the forces which molded the human brain and which overrode the older genetic programming, according to this theory.

Robert Ardrey, paleoanthropologist and popular writer, asserts quite a different theory. According to Ardrey, Lionel Tiger, Konrad Lorenz, and others, the human story started more than three million years ago, as our forest habitat gave way to the African savanna. On the plains, small creatures (about the size of twelve-year-old boys) were faced with a variety of fierce competitors, from sabertooths to venomous snakes. How did those clawless, puny apes, fresh out of the trees, manage to compete, survive, and eventually dominate? The *hunting hypothesis* proposes that hunting tactics necessitated cooperation, communication, and the use of tools. Therefore, our accelerated evolution was due to our own carnivority and that of our ancient enemies. We cannot discount the continued grip of those desperate, violent, hungry eons in our haste to prove the eminence of culture, say Ardrey and his colleagues.

The most provocative conversational aspect of the hunting hypothesis is the social implication that aggression is an innate trait, and the political implication that hopes of attaining peaceful cooperation are biologically doomed to failure. How do you deal with someone who uses this theory as an argument for the innately murderous nature of present-day humanity?

One way is to cite the new evidence and new theory presented by Glynn Isaac in a recent issue of *Scientific American.* His idea is that a shift from individual foraging to *food-sharing* around two million years ago could be the mystery factor which boosted our brain size. It isn't wholly our tendency to tinker with tools, or even our lust for blood combat, that made us the brainy creatures we are today. New skills, more advanced crafts, and complex social/economic behaviors could have been necessitated by a food-sharing society. *Eating,* say the food-sharing hypothesists, was the mother of consciousness; basket-weaving, not spear-sharpening, was the most truly proto-human technology.

RELATED TOPICS:

Altruism/Selfish Gene, Consciousness, DNA, Origin of Life, Sociobiology.

FURTHER READING:

"The Evolution of Man," Sherwood C. Washburn, *Scientific American,* September, 1978.

"Bushes and Ladders in Human Evolution," Stephen Jay Gould, *Ever Since Darwin,* Norton, 1977.

The Hunting Hypothesis: A Personal Conclusion Concerning the Evolutionary Nature of Man, Robert Ardrey, Atheneum, 1976.

"The Food-Sharing Behavior of Protohuman Hominids," Glynn Isaac, *Scientific American,* April, 1978.

IMAGE ENHANCEMENT

DEFINITION:

Computer improvement of digitized electromagnetic information, applied to radio astronomy, meteorology, or medical diagnostic technology.

WHAT IT REALLY MEANS:

Due to our long, danger-fraught evolution, one thing we humans do very well is *see.* However, our instinctive desire to peer into the dark, to seek out danger and wonder and hard data, to see more and better, has led us to invent the tools of observational science in order to extend our natural vision. A few centuries go, the telescope and microscope zoomed our collective view out through the vastness of the cosmos and shrank it down to the microcosms of our cells. In the last twenty years, another exciting way of seeing previously unseeable realms has developed: Computer techniques applied to visual images are materializing heretofore invisible information. This radical extension of our vision is already profoundly affecting science and our daily lives, enabling us to sense such interstellar esoterica as black holes and aiding us in diagnosing cancer or forecasting hurricanes.

Our present visual physiology reflects the survival needs of our ancestors. Because our earliest forebears were diurnal, lived in a colorful and dangerous environment, and were neither swift nor strong, today we see better in color than in black and white, we perceive moving objects more readily than static background details, and our brains have the ingrained habit of seeking patterns wherever our eyes look. We are all born with quite a sophisticated image-processing system embedded in our skulls. The lens of the eye focuses light on the rods and cones of the retina, where a photochemical reaction converts the light energy of the image into a series of nerve impulses. The nerve impulses are reconstructed in the form of a picture by the visual cortex of the brain. Physiologically, it is not the eye, but the *brain* which "sees."

When NASA planned interplanetary probes, they started to devise different ways of transmitting images. At the same time, medical diagnosticians were looking for better ways to see inside the body. The use of scanners and computer image-processing revolutionized astronomy, medicine, espionage, and meteorology, and new uses for this technology are emerging every few months. The key to this human/machine hybrid kind of sight is the conversion of electromagnetic radiation—whether it is from the narrow band of visible colors or from the far ends of the X-ray range—into *digital* form.

Instead of using a lens to focus light, the interplanetary probes were designed to sense via electronic scanners, which made a series of individual measurements of radiation intensity and translated the results into strings of binary numbers. The code for the scan pattern was transmitted across millions of miles of space as a series of radio pulses; when the signals were received on earth, a computer converted them into human-readable images on video screens.

We've all seen those pictures of outer space in the newspapers or on television. In the early days of the art, you could always recognize computer images because they were composed of little square blocks of black, white, and gray. The little squares are called *pixels,* and they are the basic picture element of all computer image enhancement systems. (Imagine that you see the world through a fine grid, like a wire mesh screen. If you looked at each individual square, one at a time, and recorded whether it was light, dark, or in between, you could reconstruct a rough image from that information.) Just getting those pixels and sending them back to mission control is a major task in itself, but once that faraway image from a space probe has been converted into numbers in a computer memory back on earth, all kinds of magical things can be done.

The computer cleans up, filters, selectively distorts, exaggerates, focuses, and censors the picture information in a variety of ways, in order to extract every possible bit of knowledge from it. The scientists who originally developed the techniques began with the relatively simple *false color* method. The computer translates information that is normally invisible—like the heat patterns over a continent, the X-ray scan of a human body, or the microwave map of a planet—into shades of color. The human eye is able to differentiate only about a dozen shades of gray but can recognize thousands of color gradations. Because humans are so good at seeing colors, we can do even more miraculous processing of the computer-created image inside our brains—like forecasting the weather or diagnosing a tumor.

False color and other computer methods of removing random noise and transmission errors, sharpening contrast, and increasing definition were originally used in interplanetary space but were quickly adopted by the weather satellite system. Today you can watch hurricane Bruce, sitting off your coast, on your own television in living false color, direct from the satellite. Satellite images and new methods of interpreting them soon broke out of meteorology into geology—they are now used to prospect for oil and minerals. More ominously, spy satellites capable of reading license plates down on earth as well as tracking missiles are the first early warning of possible nuclear attack; on the hopeful side, the increasing sophistication of spy satellites makes it possible to monitor arms limitation agreements.

The wonders of outer space and practical matters like missile silos are not the only images of interest to contemporary earthlings. In order to better diagnose and treat disease, medical doctors have long sought ways to see into the hidden micro-universes of the human body. The X ray was the first and most famous, but computer image enhancement and radiological innovations have combined to create a startling generation of diagnostic imaging systems: CAT scans, ultrasonography, radionucleide imaging, and positron emission tomography are but a few of the exotically named imaging systems to be found in hospitals today.

The names of the devices may be forbidding, but diagnostic imaging is essentially the same method used in space probes. First, the imaging system sends signals into the body (X rays, sound waves, radioisotope radiation). Then a camera, scanner, or sensitive crystal is used to detect the signals *returning* from the body (which are absorbed, reflected, or transmitted by hidden structures). The information is digitized, computer-constructed, and displayed as a numerical readout or as an image on a video screen.

The CAT scan (for Computerized Transverse Axial Tomography) is the most well-known diagnostic imaging system. In a CAT-scan examination, the patient lies on a table and is slowly moved through an opening in a large metal doughnut. An X-ray beam of low intensity rotates 180 degrees around the body, and detectors on the other side of the body feed the signal to the computer; the X-ray beam is moved a few centimeters at a time, taking X-ray "slices" which the computer puts together into a picture.

CONVERSATIONAL TACTICS:

CAT scans, because of their expense and their exotic technology, have become somewhat chic. When someone mentions that he had a CAT scan, your only counter, short of being drawn into one of those boring discussions of someone else's symptoms, is to mention one or two diagnostic imaging systems that are even *more* exotic and expensive. The Dynamic Spatial Reconstructor (DSR), for example, was recently developed at the Mayo Clinic; it cost over five million dollars to build, and it is designed to put the CAT scan to shame. The CAT scan can only take static, cross-sectional image slices about ¼ inch thick, while the DSR can slice down to 1/50 of an inch and scan thousands of times faster than CAT, allowing it to reconstruct a *moving*, pseudo-three-dimensional image. It is scheduled for full-scale operation in 1982.

Both the DSR and the CAT scanner use conventional X rays to generate the images. If you really want to get esoteric, you can talk about PETT—Positron Emission Transaxial Tomography. Instead of X rays, a PETT scanner uses antimatter produced by a particle accelerator. An entity that was only a physics theory and a plot in science-fiction stories a few years ago—the positron or antielectron—is now being used to obtain diagnostic data.

RELATED TOPICS:

Digital Computer, Fiber Optics, Holography, Information, Radio Astronomy.

FURTHER READING:

"Eyes of Medicine," Claire Warga, *Science Digest,* Spring, 1980.
The Radiant Universe—Images from Space, Michael Marten and John Chesterman, Macmillan, 1980.

INFORMATION

DEFINITION:

Any data that have been organized and communicated.

WHAT IT REALLY MEANS:

In a very real sense, survival depends on receiving accurate information about the environment. In fact, cybernetician Norbert Wiener once suggested that the world "may be viewed as a myriad of To Whom It May Concern messages." This reliance on information is as old as the first living organism, but it has been elevated to the level of a new science because of two recent discoveries: first, that information may be treated as a measurable, physical quantity; second, that new technologies, principally computers and satellites, make the storage and transmittal of information efficient. The conjunction of these discoveries has led to the "information explosion."

The mathematical modeling of information was achieved by Claude Shannon not long after the end of World War II. Although information theory involves complex mathematics, the key insights can be understood by anyone who remembers a little about probability. Suppose there are N likely events (it's either a boy or a girl); then the probability of each occurrence is 1/N (in this case, ½ or 50-50). When we are told which event actually has occurred (it's a boy), we have acquired some information, and our uncertainty has been reduced. Furthermore, the larger the number of likely events (boy, girl, twins, triplets), the more information we possess when we finally find out exactly which event did occur, because we had less idea in advance as to the actual outcome. This means that one condition of any measure of information is that it increases with N—the number of possible outcomes. Think of it this way: telling you who won a horse race gives you more information than telling you who won a boxing match.

The second important condition is *additivity*: the sum of the measure of two pieces of information should be the same as the measurement of a single piece of information which specifies exactly the same outcome. If I tell you it's cold out and also tell you it's raining, my message has the same information content as telling you it's cold and raining out. In itself, additivity sounds too obvious, too trivial, to be of any use; Shannon's insight was his perception that these two common-sense criteria could be mathematically modeled using the logarithm function.

The heart of information theory is the equivalence between the key information statement (the event is the Mth of N alternatives) with the function log of N. What made this theory important in our lives was that when N is equal to 2, the unit of information is known as a *bit* (binary digit), a form of data which was perfectly compatible with the circuitry of digital computers.

At the level of binary digits—the mathematical realm of information as computers use it—any piece of information can be specified by asking yes or no questions, which can in turn be modeled by turning switching elements on or off. The universe, to a computer, can be described by playing an extended game of twenty questions. Suppose you want to describe the exact location of something on a map. The first question would divide the domain into two parts: Is it in the left half? Whether the answer is yes or no, you have reduced uncertainty by one half. The next question—Is it in the top quadrant or the bottom quadrant?—

halves the uncertainty again, which means you have quadrupled the amount of information. Sooner or later, through a process of eliminating alternatives, you will specify the exact location of the target.

The same procedure, applied to numbers, produces the binary system. Suppose you want to specify the number 583. The first question would be: Are there any 1s? The answer in this case is yes, so the first digit is 1. The next question is: Are there any 2s? And the following questions follow the logarithmic function of N equals two: Any 4s, 16s, 32s, 64s, etc? By convention, a 1 equals a yes and a zero equals a no, and binary numbers are read from right to left, so 583 would be represented by 1001000111. By looking at the digits, and recreating the series of yes-no questions, you can determine that there is one unit of 512, one unit of 64, one unit of 4, one unit of 2, and one unit of 1. It takes ten binary digits, ten bits, to represent 583.

The theoretical elegance of information theory was matched by powerful, emerging computer and telecommunications technologies. The computer's ability to manipulate and process information was probably the precipitating cause of the information revolution, but simply having the capacity to handle information would be uneventful if we did not have a way to communicate it. The telephone lines, microwave stations, satellites, and frequency spectrum are the information analogs to the transportation grids of the industrial economy. Whereas mobility in physical space is achieved through roads and railways, mobility through information space is gained through the telecommunications network.

One crucial link in that network is the geostationary satellite. In 1970, there were only 14 such satellites. By 1980, there were 110, each with greater communication capacity, more effective radiated power, and a hundredfold decrease in costs over the original ones. The early 1980s should see such advances as even larger satellites with greater on-board power generation and with on-board transmission-processing capabilities, multiple spot satellite communications beams, and more sophisticated access and channel-sharing schemes. These advances should lead to direct satellite communication between individual business establishments.

The hottest area of communications technology is the direct joining of computers into the network. *Digital encoding* and transmission of signals (the translation of information into bits, and bits into electronic pulses) allows remote computers to be interconnected over compatible communications channels. This allows digital *error correction* to preserve accuracy of information transmitted over long distances; *burst multiplexing* and switching increases the efficiency of communications channel use by allowing single channels to be shared among multiple data sources and resources; *packet switching* allows the data that will be transmitted to be partitioned into small "packets" that can then carry their own destination address and find their own independent paths through a communications network.

CONVERSATIONAL TACTICS:

The social impact of the information explosion has been and will continue to be deep and broad. When it gets personal, information isn't always seen as a beneficial commodity. Take, for instance, the fight over data banks and information privacy.

Personal documentation privacy refers to those standards or principles that promote fairness to individuals in the use of recorded information about them. This means that it concerns rules governing the third party *disclosure* of personal data; rules governing the *collection* of personal data and the subject's *access* and *corrections* rights regarding the data; and standards for the maintenance, management, and use of personal data. In short, who is entitled to know what about us?

Everyone has a story about a neighbor who couldn't get a loan because the bank heard from the insurance company that the car had been in two accidents this year, or the colleague at work who couldn't get an American Express card because the BankAmericard computer insisted (incorrectly) that a large sum was owed, or the lady who couldn't collect Social Security because the computer insisted she was dead. Life can get complicated when those strings of electronic pulses encode details about individuals' lives. Before these horror stories happen to you, *Talking Tech* wants to apprise you of your information-privacy rights. The federal government and more than a dozen states have now adopted legislation which embraces the following principles enunciated by HEW's Committee on Automated Personal Data Systems:

1. There must be no personal data record-keeping systems whose very existence is secret.

2. There must be a way for an individual to find out what information about him/her is in a record, and how it is used.

3. There must be a way for an individual to prevent information about

him/her that was obtained for one purpose from being used or made available for other purposes without consent.

4. There must be a way for an individual to correct or amend a record of identifiable information about him/her.

5. Any organization creating, maintaining, using, or disseminating records of identifiable personal data must assure the reliability of the data for their intended use and must take precautions to prevent misuse of the data.

These five principles form the core of the strategy of subject participation. Should you decide you don't want to participate, remember the line attributed to Justice Brandeis: An individual's right to be left alone is the most valued right of civilized people.

RELATED TOPICS:

Analog Computer, Digital Computer, DNA, Entropy, Fiber Optics, Mathematical Modeling, Miniaturization, Randomness.

FURTHER READING:

Data Banks in a Free Society: Computers, Record-keeping, and Privacy, National Academy of Sciences, Quadrangle Books, 1972.

From Spark to Satellite, Stanley Leinwoll, Scribner's, 1979.

The Coming of Post-Industrial Society, Daniel Bell, Harper & Row, 1973.

INTELLIGENCE ENHANCEMENT

DEFINITION:

The use of pharmaceutical, electronic, and behavioral means to increase the level of human intelligence.

WHAT IT REALLY MEANS:

Witlessness is obsolescent. Just as the primeval forest served as an evolutionary filter, weeding out those maladaptive types who couldn't use their thumbs to swing from branch to branch, the complexity of modern life is weeding out those who aren't swift enough to play in traffic. The world has become way too weird for anybody to make it on dumb luck alone. Either we get smarter as we go along, or our daily lives will become unmanageable.

While one school of psychology fights over the meaning of IQ tests, and another branch debates whether intelligence is 80 percent genetic and 20 percent environmental (or vice versa), a group of technologists from diverse scientific disciplines are making an end run around the whole matter. Today, because of breakthroughs in a dozen new sciences, a different view of human intelligence is gaining ground. If these new theories are correct, the limits of intellect are not indelibly stamped in our brains at the moment of conception, and everybody's IQs may be much more elastic than previously believed. Biochemical, electronic-cybernetic, and behavioral approaches to intelligence expansion are now being explored; early evidence indicates that such tools can be used to improve memory, boost learning and problem-solving ability, and stimulate imagination.

Nobody has yet marketed an IQ milk shake, and we're certainly not suggesting that you can run out to your local pharmacy and order a dozen units of intelligence booster, but there are some early research results in psychopharmacology which bear close watching. Consider the case of the geriatric antidepressants. Loss of memory due to aging is one area where intelligence-expanding drugs have a legitimate place in the medicine cabinet. Pharmaceutical chemists have created a variety of drugs for this purpose, and some of the new drugs may have the potential for increasing memory, concentration, and problem-solving abilities in nonsenile subjects.

One particularly intriguing drug of this type is *Hydergine*. (Ask your friendly family physician to read "Workshop on Advances in Experimental Pharmacology of Hydergine" in *Gerontology*, 24 [1978].) The chemical structure of Hydergine is similar to that of LSD, but its action is quite different from its hallucinogenic relative. Hydergine is approved by the FDA for treatment of memory loss and depression in aging patients. It has been suggested that the drug stimulates production of a natural brain substance which can grow new nerve tissue.

Another suspected intelligence-enhancer is *vasopressin*, which is available in synthetic form as a prescription nasal decongestant. Vasopressin is also a naturally occurring chemical messenger that regulates water balance and other important functions in the body but which acts in the brain as a facilitator of memory and learning. Commercially available in the form of *Diapid* nasal spray, synthetic vasopressin has been studied by research groups in the U.S. and Europe as a means of restoring memory to amnesia victims and as a way of improving the attention span of aging patients.

Acetylcholine is one of the most important chemical transmitters in the brain, and the level of acetylcholine in the brain appears to affect learning ability. A simple way to adjust your level of acetylcholine is to increase your intake of *choline*, the raw material your body uses to construct it. Choline is found naturally in meat, eggs, and fish (there might be something to the old "fish is brain food" myth, after all). It can also be found in *lecithin*, a harmless nutrient supplement which can be obtained in any health-food store.

Intelligence is a blend of many abilities. Attention, concentration, imagination, motivation, memory, perception, organization, and communication skills are all involved in measurable intelligence. Drugs that act on some of these aspects already exist, and it looks like new IQ uppers will be invented every year. According to the statistics, many of the people you see at home or on the street are likely to be taking tranquilizers such as Valium or Librium. Is it any less moral or practical to take drugs to help find solutions to your problems than to take them to escape the problems?

We may live in a pill-oriented society, but it is also a high-tech society, and some recent studies indicate we may be able to bypass drugs entirely by entering into a mental bootstrapping relationship with our *machines*. The scenario: Humans devise machines of low intelligence but with special talents like high-speed information processing or audio-visual modeling, devices which can then be used to help us become more intelligent; with this higher intelligence, we can create machines of greater intelligence, which will enable us to become even more intelligent. The key to this possible future lies in the human/machine relationship.

According to a recent study on computer-assisted learning, it is now possible to use a computer, a video screen, and audio speakers as a kind of Nautilus machine to exercise the mind. It works by forcing the brain to integrate information through both hemispheres at the same time.

Through "dual coding"—putting information in both verbal and visual languages—the words, sounds, and images naturally lead the separate parts of the brain into a coordinated effort. With this type of learning device, you don't just accumulate knowledge, you actually get smarter automatically as you handle more and more information.

Biofeedback, once touted as "electronic Zen," has developed into another potential intelligence tool. Some investigators think that a type of feedback known as *cross-correlation* can also show people how to coordinate the different parts of their brains. People can learn to increase their own interhemispheric efficiency in the same way they can learn to control body temperature, blood pressure, or their brain-wave frequency. Instead of beeping when your brain wave hits the specified frequency, a feedback correlator would signal you only when two or more portions of your brain are acting in synchrony.

Metaphysics may seem like a strange topic for a book about science and technology, yet the last few years have seen an increased understanding of the meditation practices of yogis and Zen masters. The biomedical research clearly indicates that the regular practice of meditation will improve your blood pressure and your pulse and respiration rates, hence increasing the flow of oxygen to your brain tissue. When your brain cells receive more oxygen, you think better. With every embellishment removed, meditation consists of nothing more mystical than sitting quietly and comfortably, adopting a passive mental attitude, and repeating a word or phrase such as "Om" or "One" in a fixed, regular rhythm. Do that for twenty minutes twice a day, and you can up your IQ without using nasal spray or electrodes.

CONVERSATIONAL TACTICS:

A deep concern for one's intelligence and a healthy interest in scientific matters have not always been considered personally ingratiating traits. In fact, a recent Presidential report found that the public regarded school science courses as "elitist" and "only for the students considering college." It seems strange that one might have to defend an abiding interest in the life of the mind, but in a society that consistently gave its highest Nielsen ratings to The Beverly Hillbillies, anything is possible. If someone catches you reading *Talking Tech*, and you feel the need to defend your curiosity, simply revert to the lowest common social denominator and explain to them that intelligence has become *sexy*.

A hundred thousand years ago, girls went for the kind of guy who could stand upright and throw sharp sticks, but sexual fashions and desirable traits have changed a bit since then. We aren't hulking around the prehistoric savanna anymore—we're weaving our way through a full-blown, information-saturated, postindustrial society. In the present period of urban confusion, the ability to think is simply more valuable as a survival skill than a thick skull, a true aim, or well-defined neck muscles. Sociobiologists assert that survival skills are the hidden reasons we choose mates—which is why intelligence is sexier than it used to be.

RELATED TOPICS:

Artificial Intelligence, Behavior Modification, Bioethics, Consciousness, Digital Computer, Endorphins, Human Origins.

FURTHER READING:

"Mind Food," Sandy Shakocius and Durk Pearson, *Omni*, 1979.
The Book of Breakthroughs. Charles Panati, Houghton Mifflin, 1980.
Supermind, Barbara B. Brown, Harper & Row, 1979.
Use Both Sides of Your Brain, Tony Buzan, E. P. Dutton, 1979.

LASERS

DEFINITION:

Any one of several devices which convert mixed-frequency electromagnetic radiation into one or more discrete frequencies of coherent, highly amplified, visible light.

WHAT IT REALLY MEANS:

It's no wonder that "laser beam" has become one of the most often repeated tech-talk mantras. Twenty years ago, neither the word nor the device existed; today, lasers have become indispensable to science, industry, communications, and entertainment; tomorrow, lasers will repattern our lives in ways we can't predict. Certainly, "laser" is one of the most potent magic words ever conjured by the wizards of science and technology, but it is also a reminder that even the most farfetched, unearthly, outright metaphysical theories about the nature of the universe may later have astounding technological applications.

Like levers and inclined planes, lasers are simple machines because they follow a fundamental law of the physical world to amplify energy. Levers and planes use the forces of gravity and inertia to amplify the power of our muscles. Lasers use a force from the subatomic level to amplify the power of radiation. The word itself is the key to the basic principles of the device: LASER is an acronym for Light Amplification through Stimulated Emission of Radiation. Taken all in a lump like that, it tends to sound complex and techno-mystical. When it is taken word by word, it becomes clearer and more accessible to the level of common experience.

Light is probably the single most important element in our lives as sentient beings, and it is undoubtedly the single most important topic in twentieth-century physics. The quantum revolution at the turn of the century set the stage for laser theory by declaring that light is both a particle (a thing) and a wave (a process). If light didn't behave in both of these apparently inconsistent ways, lasers wouldn't be possible; a laser beam is a literal materialization of quantum theory.

The tremendous power and sharp focus of laser light is a result of its coherence: A beam of coherent light consists of photons which vibrate at the same frequency, wavelength, and amplitude. Because light behaves like a wave, the result of aligning all the photon waves in phase with one another has the same result as one ocean wave overtaking another—the waves *amplify* each other to produce an even stronger wave. But that happens a billionth of a second later in the story.

The quantum aspect of light which takes advantage of its particlelike properties is in the *stimulated emission of radiation* part of the magic word. Although it comes at the end of the acronym, it actually happens at the beginning of the laser burst. Just as levers use a feature of gravity to accomplish work, lasers use a feature of subatomic forces. Quantum mechanics states that energy is not emitted or absorbed by atoms in an unbroken flow; radiation must be *quantized*—broken up into separate packets. Because of this restriction, atoms can move between different energy states only in discrete jumps in energy levels that correspond to the addition or subtraction of a particle. Lasers take advantage of this restriction to turn incoherent, low-energy radiation into coherent, high-energy light.

The energy level of an atom is materialized in the orbital clouds of electrons surrounding the nucleus; when energy is added to an atom, an electron absorbs a photon and jumps to a higher-level orbit. When an atom *loses* energy, an electron falls to a lower energy cloud and emits a photon. Because of another basic physical law, the thermodynamic concept of entropy, one of the rules of the game is that high-energy-level atoms tend to naturally decay to lower energy states, but not the other way around. Light originates in the heart of atoms and travels through space to collide with our eyes. The colors we perceive are the ways our brains translate the frequency of the photons into sensations.

Lasers use these peculiarities in atomic behavior by raising a substantial number of electrons in a population of atoms to a higher energy orbit; when the atoms inevitably return to a lower energy state, they will emit light at precise frequencies. This is the "stimulated emission." The lasing material is first "pumped" by a powerful flash of light. When the electrons in the pumped-up population of atoms fall into lower, more stable orbits and emit photons, the amplification part of the process begins.

The laser material—a gas, liquid, or crystal—is enclosed in a cylinder with half-silvered mirrors at both ends. As photons of the same frequency are emitted into the laser cavity, they align themselves in phase as they bounce back and forth between the mirrors. This increases the energy of the growing beam. As the photons travel back and forth, they pass through the pumped-up laser medium again and again, stimulating avalanches of more and more radiation in a kind of chain reaction. When the beam attains a high enough energy level, it penetrates the partially silvered end of the mirror array. The light that comes out is far more powerful than the light which went in, and there you have it: Light Amplification through Stimulated Emission of Radiation.

The properties of coherent light have made lasers unbelievably adaptable and useful tools in every corner of contemporary science and technology. In an era in which "space age" is already an archaic phrase, the image of the laser still evokes awe and wonder. We have barely glimpsed the first dawning rays of the Laser Age, although its theoretical foundations were constructed nearly a century ago, and its total impact on our world won't reach full power for another ten years.

CONVERSATIONAL TACTICS:

Lasers are a conversational magic carpet, because you can ride them to any scientific zone you want to discuss. Is the price of oil and the future of energy resources a focus of debate? Lasers may be the means by which enough heat is generated to initiate controlled thermonuclear fusion—the much-needed energy source of the twenty-first century. Is medicine the current topic? Lasers have been called "bloodless scalpels" because they automatically cauterize tiny blood vessels; they can also cut a sharper slice than any knife made of steel. Delicate eye surgery and neurosurgery are now possible for the first time.

The use of lasers to cut, weld, drill, and measure precision components has started a virtual mini-industrial revolution. Just as computers invaded business and factories a decade ago, lasers have now infiltrated every type of light and heavy industry—even the microcircuits of the newest computers are designed and etched with laser beams. The properties of microscopic semiconductor lasers make them ideal electronic-photonic transmission media for the coming fiber-optics global communications network. Tunable gas lasers (with adjustable output frequencies) are proving to be phenomenally useful in chemistry, picking rare trace elements out of mixtures and even directing the synthesis of new compounds according to the molecular architecture selected by chemists.

Geologists use lasers to keep track of movement in fault zones. A laser bounced off a reflector array that the *Apollo* team left on the moon was used to estimate the weight of the earth—6.586 billion trillion tons! Lasers are heavily studied by the Defense Department—laser-aimed ballistics ("smart bombs") are state-of-the-art weaponry, and the war-tech boys can't let the old dream of a death ray get away. New ways of detecting tumors, treating burns, aligning and surveying land, and monitoring manufacturing processes have also been developed. Laser space drives—a form of energy used to propel a space vehicle—have been designed, laser art forms have been displayed in museums and across entire cityscapes, and laser-produced holograms open a door to the world of three-dimensional imagery. Take your pick of conversational pathways—from communications revolutions to futuristic art forms to techno-medicine—and you'll find that lasers can get you where you want to go.

RELATED TOPICS:

Blackbody Radiation, Energy Alternatives, Fiber Optics, Fusion, Holography, Miniaturization, Quantum Theory.

FURTHER READING:

The Techno/Peasant Survival Manual, Colette Dowling, Bantam, 1980.
Lasers, the Miracle Light, Larry Kettelkamp, William Morrow, 1979.

LONGEVITY—IMMORTALITY

DEFINITION:

The application of scientific and technological knowledge to the prolongation of human life beyond the current, statistically normal span.

WHAT IT REALLY MEANS:

What is death? Can the aging process be retarded or reversed? How far can the individual human life span be extended? As a result of breakthroughs in nearly every sci-tech field, the means to answer those ancient mortal questions have finally become available. The full-scale research offensive on life's ultimate mystery involves a coalition of different disciplines, from biology, pharmacology, and therapeutics to bioengineering and communication theory. The most conservative interpretation of the early evidence is that life spans are *already* extendable far beyond the biblical allotment of threescore and ten, and the near-future goal of multi-century life spans is a serious consideration in life-extension circles.

With artificial hearts, pharmaceutically rejuvenated cells, cloned organs, and doctored DNA, a substantial segment of the population of people reading this book today may live to toast New Year's Eve in 2100 A.D. Life-extension technology is presenting thorny problems for bioethicists and will sooner or later have profound political impact, but the progress thus far achieved poses an unprecedented challenge to every person alive today. If you live long enough to arrest your aging process, you could be one of the first generation of near-immortals; if you miss the boat and die of one of the aging-related diseases, you may be part of the last generation of short-lived mortals.

Although the longevity quest encompasses many thousands of separate research and development efforts, the overall offensive can be grouped into three fronts, each attacking death and aging at a different level of analysis. The molecular biologists are searching on the *cellular* and subcellular plane, reasoning that if the DNA code holds the secrets of life, perhaps it also encodes the mysteries of death. On a more familiar level, the world of medical research and treatment, physicians and researchers are concerned with all the ways organ systems can go wrong; attention is presently most sharply focused on the *immune system,* in hopes of reinforcing the body's own defenses against the ravages of time. The third front—*bioengineering*—regards the body as a kind of machine, a complex and superbly engineered mechanism which simply tends to malfunction with age. The development of artificial replacement parts for vital human components has proceeded at a speed that lately has outpaced science fiction.

Cellular Theories of Aging

You remain the same person throughout your life, but your cells grow and change and die many times. If cells didn't know when to die, you would never have developed from a single fertilized cell, inasmuch as the contours of your body were shaped before your birth by the deaths of certain cells and abundant growth of others. As far as your life span is concerned, aging and death are not expressed in individual cells but in cell lineages, in generation after generation of cell populations. Is aging written into our cells?

Biological theorists currently agree that the "blueprint" governing physical changes associated with aging is likely to be found in the genetic code. Exactly how aging and death are triggered or allowed to happen by the DNA control center in each cell is still unknown, but three main hypotheses have emerged. In the first scenario, the manifestations of senescence are the result of a pretimed program lodged in the chromosomes—specific "aging genes" that give the protein-coded message to trigger menopause, grow gray hair, or thicken artery walls when the programmed time arrives. If an aging gene can be identified, a way might be found to neutralize it.

A second theory proposes a finite pool of genetic information in the chromosomes, a limited number of molecular messages available to direct the biochemical processes of life. Every organism sooner or later uses up all the stored information it needs to keep renewing cells, repairing damage, and efficiently performing vital functions. Specific aging genes are not required in this scheme, and there is some support in the observation that much of the DNA in human chromosomes appears to be redundant, as if it were being held in reserve. If this hypothesis is confirmed, it might be possible to retard the process by cloning each individual's DNA as a "cellular information bank" against the day the inborn mortal fuel is depleted.

The third hypothesis of cell aging, known as "the error catastrophe," is easily recognizable to anyone who has ever used a copying machine. Copy a page of print, then make a copy of the copy. The image gets fuzzier with each copy. If you continue the serial copying sequence, the page inevitably becomes unreadable. The DNA in your cells resembles that copying process in the sense that the "page of instructions" needed to specify the proteins of life is recopied serially every time your DNA replicates. However, unlike the copied page, every time the DNA code is copied, there is a chance for faulty information to enter the instructions. When enough errors enter the system, the bio-informational copying mistakes manifest themselves as aging. When a threshold of errors is reached, the cell can no longer reproduce itself because it is copying as many mistakes as correct information.

This theory fits remarkably well with the thermodynamic and cybernetic concept of entropy: Everything in the universe, even life, is susceptible to a buildup of errors, simply because it is easier for mistakes to enter systems spontaneously than it is for mistakes to spontaneously correct themselves. It is also known that genes have built-in mechanisms to repair damage from radiation, chemicals, and viruses—chemical mistakes that lead to the propagation of informational errors. If senescence originates in the breakdown of the error-correction system, death might be a matter for communications engineering at the cellular level. Present-day knowledge of error-correction systems, which grew out of the space program, may someday be applied to inner space in the hope that DNA can be altered to filter out the "error" of aging.

IMMUNOENGINEERING

When people age, fall ill, and die, it is usually not because of the failure of individual cells but because something goes wrong at the level of the huge populations of cells which make up the body's organs. To the longevity researcher, the most exciting physiology is the complex of cells and organs that keeps every healthy person from succumbing to the attacks of a trillion microscopic invaders, the *immune system.* Immunology is a relatively young science, but there are already subspecialties and sub-subspecialties. In the broadest sense, any progress made on ways to bolster these defenses means considerable gains against the diseases of aging.

Most specifically studied as a clue to longevity is that long-mysterious centerpiece of the immune system, the *thymus* gland. Aging is intimately related to the decline of the immune system and to the progressive shrinkage of the thymus. The warriors of your bloodstream are known as T lymphocytes or T cells; their task is to kill all invaders from outside and to protect the body from its own cancerous insurrections. T cells are produced in the bone marrow, but they must mature either in the thymus or as a result of a hormone secreted by the thymus. This long piece of tissue, located behind the breastbone, is largest early in life and starts shrinking after puberty, slowly atrophying over a lifetime. Consequently, T-cell production slows with age. Eventually, external infections or internal malfunctions get the best of the dwindling T cells. To what degree is aging a progressive failure to defend ourselves? What is it that the thymus uses up as we age, and can it be replaced?

Dr. Takashi Makinodan and colleagues at the National Institute of Aging transplanted thymuses and bone marrow from young mice into older ones, and the disease-fighting capacity of the older mice was restored to the level of animals one fifth their age. The rejuvenated mice exhibited high immunity to infection and lived over one third longer than their normal expected life spans. Makinodan compared this to a human rejuvenation of twenty years. The secretions of the thymus gland, specifically *thymosine,* are also being investigated, because thymosine levels also drop as people age. When the function of the thymus in aging is more clearly understood, newly developed biological techniques such as cloning, regeneration, and recombinant DNA engineering may enable us to grow our own personal "cell banks" while we are young, in order to bolster and rejuvenate our immune systems in our second and third centuries.

Bioengineering

A human body may be greater than the sum of its parts, but when one of its vital parts fails, the whole body dies. Bioengineering is the high technology and practical art of prolonging or augmenting life by replacing failed organs with artificial parts. The spin-offs from the arms race and aerospace programs—miniaturization, new materials, computer advances—have spurred the creation of a bionic research and development program that already dwarfs the archaic television counterpart, the six-million-dollar man.

Long before the bionic hardware boom of the sixties and seventies, a dedicated Dutch physician became the first modern bioengineer when he invented the technique of kidney dialysis in 1943. Working under near-secret conditions in Nazi-occupied Holland, a young Dr. Johan Willem Kolff succeeded in artificially duplicating the blood-filtering function of the human kidney. Dr. Kolff made more than a medical breakthrough with his first crude contraptions of cellophane and salt water—he created a new alternative in longevity research. Dr. Kolff is still working to repair nature's malfunctions with human ingenuity. Now at the University of Utah Medical Center, he supervises the world's most comprehensive bioengineering team: electrical engineers, surgeons, physiologists, mechanical engineers, computer scientists, immunologists, mathematicians, chemists, biologists, and metallurgists work together toward prolonging life.

Plans, prototypes, and working models now exist for artificial hearts, bones, eyes, pancreases, tendons, lungs, intestines, arms, livers, arteries, valves, joints, ears, skin, blood, and bones. When these devices are thoroughly tested on animals, they will be tried on humans who are desperately afflicted. If they prove effective, they will become publicly available, just as cardiac pacemakers and kidney dialysis machines became available after they were tested on patients who had estimated life spans of hours. Thanks to recent theoretical and clinical breakthroughs, the artificial eye and ear are closer to reality but still require years of further development. Fully functional artificial blood may take as long, but at present there are many bionic replacements which are closer to public availability.

The wearable artificial kidney, designed by Dr. Stephen Jacobsen, a mechanical engineer who runs the Projects and Design laboratory for the Utah facility, is a cheaper, eight-pound version of the expensive, bulky dialysis machines devised by his colleague, Dr. Kolff. The ultimate refinement will be to eventually build a microminiature implantable artificial kidney. The Utah arm, also designed by Dr. Jacobsen, is truly bionic because amputees are able to move their prostheses by merely thinking about the movement they wish to accomplish. By linking microprocessor-guided artificial muscles made of plastic fibers to the stumps of neuromuscular trunks, Dr. Jacobsen's team put the patient's own brain in charge of the new arm.

Others at Utah are working on equally vital goals. Dr. William H. Dobelle started the artificial vision program in 1969, and cortical stimulation experiments have made possible the first crude attempts at simulating vision. Eventually, miniature television cameras mounted in glasses or implanted into the eye sockets may translate information to the visual cortex. The artificial heart, certainly one of the greatest hopes of a true bionic longevity program, is under development by Robert Jarvik's team. Molded polyurethane hearts have been implanted in calves, and if a working human model is finally perfected we may all replace our hearts every sixty or seventy years.

CONVERSATIONAL TACTICS:

If you have a conceptual grasp of cellular aging, immunoengineering, and bioengineering, you should be able to talk life extension with aplomb. This does not guarantee, however, that you will know any immediately useful information about prolonging your own life. Never fear, teams of scientists have been looking at statistics and life-styles, attempting to discover what long-lived folks do that other people don't. If you want to add as many as fifteen years to your own life span without resorting to longevity potions or bionic organs, it may be a matter of following a few simple guidelines in your daily life.

A recent study of seven thousand people in Alameda County, California, found that men could add eleven years to their lives, and women could add seven, by following seven rules: Drink moderately; don't smoke; eat breakfast; eat at regular times; exercise; stay slim; and get eight hours of sleep each night. Parallel research in other areas has augmented this study. If two other additions are made to this list—avoid excess salt, and take twenty minutes each day to nap, relax, or meditate—it is probable that more years would be added.

One odd corner of the longevity world worth conversing about is the controversial role of food preservatives, because it is a rare reversal of the usual role of synthetic chemicals in our food. While other food additives are being denounced as carcinogens, a class of chemicals known as *antioxidants* may be prolonging our lives while they increase the shelf life of our food! The most common antioxidant, BHT, is added to cereal, margarine, potato chips, and cookies. Studies indicate that rats fed BHT live up to 20 percent longer. Theoretically, these chemicals work their magic by cleaning the cells of dangerously reactive molecules known as "free radicals." If this research bears conclusive results, we may find that some of the garbage in the food we eat is counteracting some of the other garbage we eat.

RELATED TOPICS:

Bioethics, Cholesterol, Clone, DNA, Entropy, Information, Mutation, Regeneration.

FURTHER READING:

The Life-Extension Revolution, Saul Paul Kent, William Morrow, 1980.
The Complete Book of Longevity, Rita Aero, G. P. Putnam's Sons, 1980.
"The Real Bionic Man," Dick Teresi, *Omni,* November, 1978.

MATHEMATICAL MODELING

DEFINITION:

Any system of equations which can describe or predict behavior in physical systems.

WHAT IT REALLY MEANS:

Twenty-five hundred years ago, a group of Greek philosophers known as Pythagoreans made the astonishing claim that "infinity itself and unity itself are the substance of the things of which they are predicated;

this is why number is the substance of all things." The idea of mathematics as more than a description of the universe—as the very basis of the cosmos—has cropped up all over the world. While few people today would be willing to claim that numbers are the ultimate constituents of reality, more than a few mathematicians have claimed that numbers are an essential vehicle for *understanding* reality. As mathematical sophistication has increased, and as computers have improved our ability to perform calculations, it has become clear that certain aspects of the physical world may be modeled mathematically.

A model is simply an analog of the real world. Like a dollhouse, a model represents the structure or processes of some aspects of a larger world in simplified form. Since we can't always experimentally manipulate the world, we manipulate the model in a simulation exercise and extrapolate our results to the real world. We are all familiar with hundreds of models: dollhouses and miniature cars are physical models, blueprints and wiring diagrams are two-dimensional models of three-dimensional structures, and budgets are mathematical models of financial behavior. The great breakthrough of our time has been the ability to develop mathematical models of complex behaviors and events and then to use the model not just to describe but also to predict.

Many of the mathematical modeling procedures have strange and unfamiliar names; all of them necessitate the use of complex mathematics. However, the essence of each subcategory of modeling may be explained in a paragraph or two:

Catastrophe theory: This theory, developed by the French mathematician René Thom during the last twenty years, is not a model of a particular aspect of the physical world; instead, it is a new way of looking at a phenomenon which all aspects of the world share—*change*. According to Thom, a catastrophe is a discontinuous transition, a jump between two stable states. Water's jump from a solid to a liquid state is a catastrophe, as are electrons' quantum jumps and bubbles bursting, and people changing their minds.

The standard mathematical model for studying change is the calculus—a quantitative method devised for dealing with smooth, continuous change like the motion of an accelerating body. Thom's theory, which deals with *abrupt* changes, is qualitative and geometrical. As incredible as it seems, he has used the method of differential topology (calculus applied to the geometry of shape) to categorize all elementary changes into seven types of curve. He believes that different kinds of change exhibit similarities—that mud cracking in the heat, or a plaster wall cracking, or the lines of a face, are all examples of one of the seven fundamental types of change. Another complex curve models phenomena as weirdly diverse as stock market behavior and "cathartic release from self-pity." Understand these seven curves and you have a map of how change occurs. If Thom is correct, all science, from physics to psychology, may be subsumed under the study of catastrophic change.

Econometric models: When the Nobel prize committee decided to begin awarding a prize in economics, it was essentially acknowledging that economics had become a quantitative discipline. Even if it was still the "dismal science," it was at least rigorously dismal. Actually, attempts to mathematically model national economies are about a hundred years old. However, it has only been in the last forty years, beginning with the publication of John von Neumann and Oscar Morgenstern's *Theory of Games and Economic Behavior*, that models of any scope and predictive power have begun to be developed. Prior to this work, most efforts to model economies dealt solely with supply and demand. John von Neumann's contribution was his recognition that "the typical problems of economic behavior are strictly identical with the mathematical notions of suitable games of strategy." This insight improved econometric models by allowing them to incorporate the real-life behaviors of producers and consumers.

These models try to do for national economies what we implicitly do as individuals with our personal microeconomies—describe economic behavior and predict how that behavior will change under the influence of new conditions. If one receives a 15 percent raise, how much of it is spent and how much saved? How does one react to the opportunity to work overtime? How much must coffee prices rise before one switches to tea? Imagine trying to pose, quantify, and answer a similar set of questions for the U.S.A's trillion-dollar economy. The first step is to develop a comprehensive list of the major influences on economic activity (things like interest rates, savings rates, and consumer spending). The next step is to generate a list of assumptions, based on past trends, regarding how the marketplace is likely to operate in the future (things like how consumers traditionally react to 5 percent more income, how businesses are known to react to tax incentives, what the government is likely to do if inflation hits a certain percentage).

Finally, the influences and assumptions are mathematically related through linear equations. For example, in the prestigious Wharton School model the general price level of manufactured goods is a function of labor cost per unit, the overall level of manufacturing production, and the estimated maximum capacity of production. The equations, which *are* the model, are computerized so that many different possible economic actions can be tested for possible outcomes.

Game theory, which also developed from von Neumann and Morgenstern's work, is an attempt to model how people behave in situations in which they are in conflict, or interaction, with other actively participating persons. The key insight of the theory is analysis of games according to two basic distinctions. The first analytic rule regards the payoff rules of the games: All games are either zero-sum or non-zero-sum games.

A *zero-sum game* is simply one in which the amount one player wins must equal the amount lost by the other player. Many children's games such as tick-tack-toe and stone-paper-scissors are zero-sum games. *Non-zero-sum games* are ones in which gains and losses need not cancel out. A classic example of such a "game" in a real-life context is what Garrett Hardin calls "the tragedy of the commons": Suppose land is held in common for purposes of animal grazing—the original purpose of fields like Boston Common. The optimal strategy for each farmer individually is to graze the maximum number of his own cattle. Yet when every farmer individually maximizes, the land becomes overgrazed and everyone loses. The move which would then benefit everybody would be the reduction of the number of cattle to the carrying capacity of the land, but it never pays any one individual to reduce stock unless all the other players also participate. The payoff rules for such a game might look like this: If everyone cooperates, they all receive moderate gains. If no one cooperates, they all receive minimal gains. If all but one player cooperates, everyone receives slightly less than moderate gains and the "cheater" maximizes his or her gain.

The second important distinction of game theory concerns the amount of information available to the game players: In a game of perfect information, each player sees the whole situation and each knows as much about it as his opponent. In a game of imperfect information, each player knows his own resources but not those of the opponent. Checkers and chess are examples of the first type, while gin and poker are examples of the second. Of course, it is possible to combine the two distinctions: Poker is a zero-sum, imperfect-information game. The ultimate utility of game theory as an analytical tool is constrained only by our ability to perceive human interaction in a gamelike setting with rules, moves, and payoffs. Such a perception goes to the very heart of one's view of human beings and human nature.

Operations research, sometimes known as decision theory or systems theory, is the application of the scientific method and quantitative techniques in order to solve problems in management science. This may sound rather dry, but OR was born in a situation of utmost urgency. In Britain during World War II, the Air Force needed a scientific way to allocate fighter planes and pilots. The U.S. Navy used OR to design and schedule antisubmarine search-and-destroy missions and to evade such measures on the part of the enemy by adopting a strategy first developed in a 1928 theoretical paper by the ubiquitous John von Neumann (he was also a key figure in the development of two other children of that war—the atomic bomb and the computer).

Today, OR is used in business and industry to solve a host of allocation problems (how can Exxon be sure that each of its service stations has enough gasoline?), inventory problems (how can GM be sure that they don't end up with too many fenders and not enough bumpers?), and routing problems (how does the traveling salesman make sure he visits all his customers in the least time for the least expense?).

One type of OR problem everyone experiences is the queuing problem: Which line do you get into at the supermarket? In the last decade, OR research has proven that the quickest way to move randomly arriving customers through a service point is by having one feeder line to the various service locations. This is the arrangement now found in most banks and fast-food services. Our ability to abstract and mathematically model all different types of phenomena—from bombing patterns to buying habits—is responsible for OR's increasingly greater applications.

CONVERSATIONAL TACTICS:

Game theory may not tell us everything we want to know about human interaction, but game theorists have developed some amusing two-person games to illustrate their theories. If you have to win at all costs, you might try "bluff," a zero-sum game with an optimizing strategy for one of the players. The game begins with you, player A, writing a number from 1 to 10 on a slip of paper. You then announce a number from 1 to 10 to player B, who must guess if the number you've announced is the one you wrote down, or whether you are bluffing. The payoff matrix for player A is:

	TRUE	BLUFF
SAME NUMBER	−$.10	+$.50
DIFFERENT NUMBER	+$.50	$1.00

Mathematical Modeling—Winning the Bluff

Suppose you wrote down "4" and said "6" (this puts you in the bluff column). If player B called your bluff, putting the decision in the *different* row, then you must pay out one dollar. Suppose you wrote "4" and said "4." Then if player B called your bluff by saying you have a different number written down, you would collect fifty cents.

Since the payoff matrices in zero-sum games are always mirror images of each other, there is a tendency to think that both players have an equal advantage. This is not always the case. In "bluff," if player A (the player writing the number) says a different number two sevenths of the time and the same number five sevenths of the time, then even if player B plays the best possible strategy (guessing true five sevenths of the time and bluffing two sevenths of the time), player A will average a return of 7.1 cents per game.

RELATED TOPICS:

Analog Computer, Artificial Intelligence, Digital Computer, Information, Randomness, Topology.

FURTHER READING:

Catastrophe Theory, Alexander Woodcock and Monte Davis, Avon, 1978.

Mathematics Today, ed. Lynn Arthur Steen, Springer-Verlag, 1978.

Tools for Thought. C. H. Waddington, Basic Books, 1977.

MELTDOWN

DEFINITION:

An event occurring in the core of a nuclear fission power plant which results in the partial or complete liquefaction of the nuclear fuel.

WHAT IT REALLY MEANS:

In order to truly understand what constitutes a meltdown, one must have a basic understanding of how a nuclear fission power plant operates. Fortunately, this is not difficult. Just keep two principles in mind. First, a nuclear power plant works in much the same way as an electric immersion coil which you might use to heat a cup of tea. Second, unlike the immersion coil, there is no cord to pull to turn off a nuclear plant. In fact, nuclear power plants have something of a life of their own. In the case of certain extreme malfunctions, that life can extend for tens and hundreds of thousands of years.

Basically, a nuclear power plant is little more than a complicated method of boiling water to generate steam, which in turn powers a turbine to produce electricity. The heat which boils the water is produced by a nuclear chain reaction, and the control (or lack of it) over this reaction is where all the complications arise.

Nuclear power plants use uranium 235 or plutonium as fuel, both of which are radioactive elements which emit neutrons. The radioactive fuel is formed into pellets which are stacked in tubes to form *fuel rods*. When the fuel rods are brought close together, enough flying neutrons

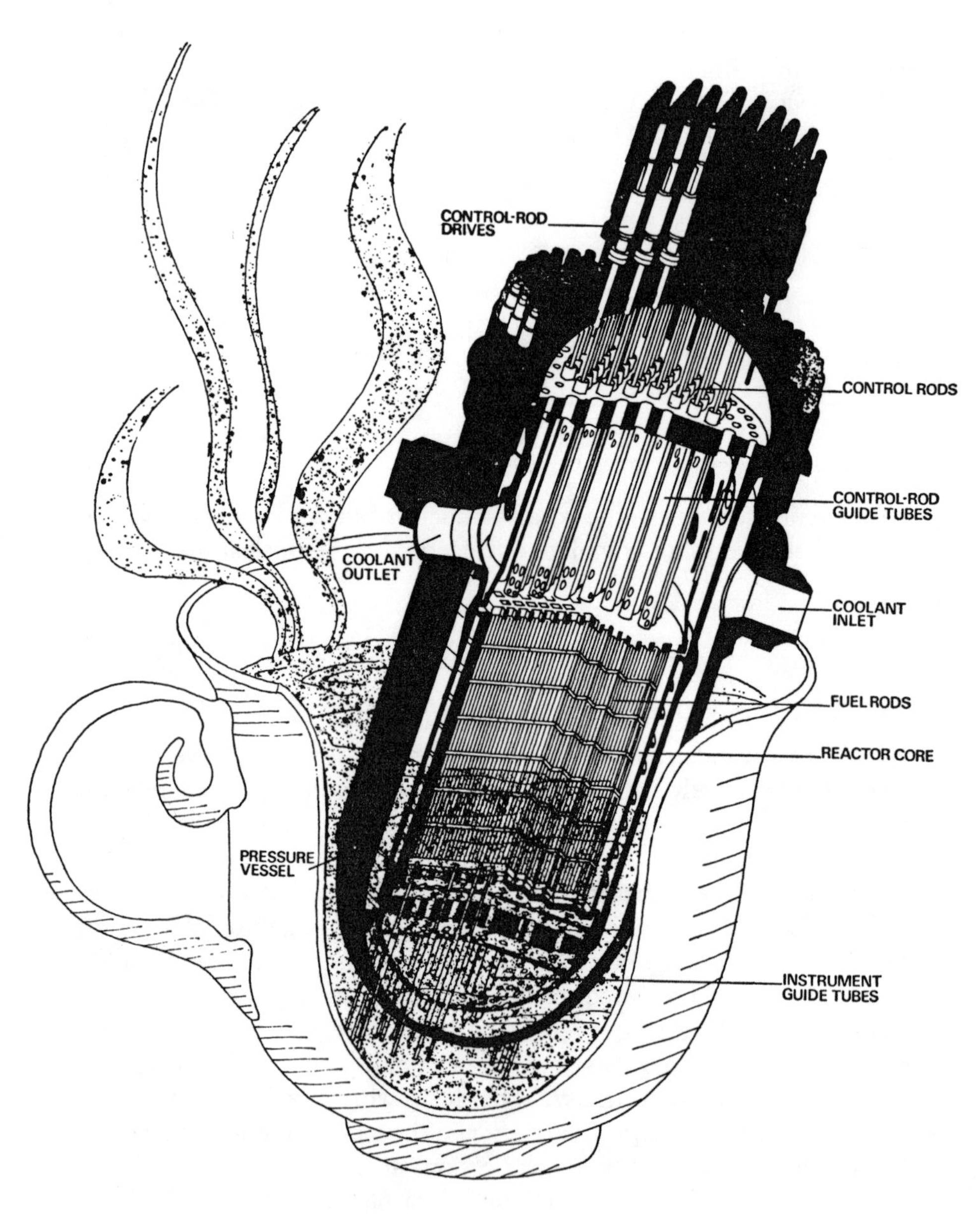

Meltdown—Heating Your Tea

will strike other neutrons in adjacent fuel rods to initiate a *chain reaction.* When one flying neutron strikes the nucleus of a fuel atom, the target atom emits *two or three* neutrons, which means that the collisions and liberated energy expand at an exponential rate.

Between the fuel rods hang rods of boron or graphite, which act as *control rods* by capturing and absorbing some of the wild neutrons without emitting neutrons of their own. By moving these control rods up and down, the speed of the nuclear reaction is regulated, and the operator may control how hot the water gets and how much power is generated. This system of rods is called the *core,* and it must be immersed in cooling water at all times. The distinctive towers often associated with nuclear plants are just structures to hold and cool the water.

Unfortunately, all of the heat being produced by the core is not under the control of the power plant operator. When people say that radioactive elements are "hot," they refer to actual thermal radiation as well as the other forms of invisible radiation. Once the control rods are pulled out of the core and the nuclear reaction begins, the core takes on a life of its own. Seven percent of the heat generated by a nuclear plant is self-generating and cannot be turned off even if the control rods are reinserted. Under normal conditions, this extra heat is simply used to produce more power; however, should there be a loss of water around the core, the temperature would quickly rise to 5000 degrees Fahrenheit.

Overheating is made more difficult to control by the fact that core radiation is extremely dangerous to living things. Invisible emissions could doom an inadequately shielded person in the core area. The temperature of an uncontrolled nuclear core is high enough to melt the mechanical devices used to handle the fuel rods and melt the radioactive fuel itself, which would in turn melt the steel support structures, the boiler, and the concrete structures beneath it. When this happens, the survivors call it a *meltdown.*

CONVERSATIONAL TACTICS:

Like all technical fields, the nuclear power industry has developed its own colorful and mystifying vocabulary. In order to really understand what is going on, you've got to deal with the lingo. Therefore, herein presented is the Everyman and Everywoman's Guide to Nuclear Power Bafflegab:

ATWS—Anticipated Transient Without Scram. What this means is big trouble. There is a malfunction in the core (a transient), and the operators cannot shut down the plant by inserting the control rods (a scram).

BLOWDOWN—A variant of the meltdown in which one of the pipes bringing water to cool the core springs a leak, causing the water to instantly escape into the air as superhot steam. This situation would leave the reactor without coolant.

CHINA SYNDROME—The worst-case scenario for a meltdown, during which the molten fuel mass (perhaps two hundred tons of liquid metal and radioactive material) would either hit an uncharted pocket of groundwater which could possibly vent radioactive steam into the atmosphere . . . or come to rest, perhaps fifty or a hundred feet below the surface. It would then sit there, remaining hot and bubbly and lethal for hundreds of years.

LOCA—Loss of Coolant Accident. The generic name for any problem which might lead to a meltdown.

SCRAM—The emergency insertion of the control rods to shut down the plant.

TRANSIENT—Any malfunction in the plant's operation, usually requiring a scram.

RELATED TOPICS:

Appropriate Technology, Energy Alternatives, Fission, Fusion, Mutation, Technology Assessment, Toxic Chemicals.

FURTHER READING:

The Silent Bomb, ed. Peter Faulkner, Random House, 1977.
The Electric War, Sheldon Novick, Sierra Club Books, 1976.
A Guidebook to Nuclear Reactors, Anthony Nero, Jr., University of California Press, 1979.

MINIATURIZATION

DEFINITION:

The development of techniques to reduce the size of electronic components; this process results in order-of-magnitude increases in complexity, flexibility, and power, accompanied by large reductions in cost and energy-use.

WHAT IT REALLY MEANS:

Every earth-dweller's life has already been irrevocably altered by the miniaturization revolution, from the sleek, microprocessor-guided missiles waiting in their launch tubes to talking lunch boxes that teach children how to spell. Over one hundred electronic pacemakers are implanted every day to regulate human hearts which would otherwise falter. Our homes, our jobs, our means of transportation, communication, and entertainment, and our essential services are increasingly dependent on these ever more potent, ever tinier devices. How did this circuitry-shrinkage process get started, why has it moved with such astonishing rapidity, what is it going to do to our lives next . . . and how *do* they make those machines so darn small?

If necessity is the mother of invention, then warfare must be the father of research and development. Just as computers were constructed to solve ballistics problems during World War II, the need to make smaller computers was dictated by the postwar arms race. A sophisticated guidance system was needed to "deliver" nuclear weapons to their targets, and it had to fit in the nose cone of a missile. Later, when NASA started thinking about humans instead of warheads in missiles, additional impetus was provided by the need for compact, reliable, lightweight life-support systems. None of these goals was possible with the vacuum-tube technology of the 1940s, so a search was initiated for a more efficient kind of electronic element. In 1947, John Bardeen,

Walter Brattain, and William Shockley invented the *transistor*, and phase one of the miniaturization cycle began in earnest.

Vacuum tubes in early computers acted as on-off switches or amplifiers, depending on how electrons were made to flow between positive and negative electrodes in a vacuum. The transistor uses the flow of electrons in purified crystals of solid silicon, which is why the descendants of the transistor are called *solid-state* technology. The special materials used in transistors are known as *semiconductors* because of crystalline properties which allow them to conduct or amplify the flow of electrons only under certain conditions. By "doping" a crystal of pure silicon with a special impurity, zones can be created where there are greater or fewer electrons—positive (P) or negative (N) layers. By slicing thin P and N crystals, and stacking the layers alternately (PNP or NPN), transistors were created. The resultant element, about the size of a pencil eraser, was not only smaller than a vacuum tube but used less power, dissipated less heat, and was far more reliable. Computers only began to grow out of their calculating-machine infancy when transistor technology was introduced to nonmilitary computers in the 1950s.

In the 1960s, the space program combined with the intense military and scientific interest in solid-state technology to initiate the second phase of the miniaturization revolution. The most simple transistor is nothing more than one P zone between two N zones, or vice versa, and *the zones themselves can be very small*, as long as they are properly separated by insulators and connected by conductors.

Scientists at Fairchild Semiconductor started to make transistors in a new way. Instead of making stacks of P and N chips, they used much smaller amounts of semiconductor, insulator, and conductor and deposited them in an integrated horizontal pattern rather than a vertical stack. They made giant drawings of how the P and N zones and various connections should be made, photographed and reduced the patterns, then used these photomasks as templates to chemically etch an entire circuit consisting of hundreds of transistor elements on an *integrated circuit* the size of a single old-style transistor.

When integrated circuitry was developed, the technology started a curious kind of inbreeding. Computers built from transistors were being used to help design integrated circuits; as computers began to be built from integrated circuits, they helped design the next level of miniaturization, known as medium-scale integrated circuitry. Other new technologies, especially the laser, made it possible to plan even more powerful and densely packed "generations" of circuitry. The number of functional electronic components which could be packed into a square inch was less than one in 1947, grew to four by 1960, increased to forty by 1962, expanded to four hundred by 1965, and exploded when the first large-scale integrated circuits were introduced in 1969, packing *forty thousand* components on a sliver of semiconductor a fraction of an inch square.

Large-scale integration (LSI) was the third phase of miniaturization. Throughout the 1960s and 1970s, older generations of computers and electronic devices were leaving military service to infiltrate factories, industries, and homes. "Space-age technology" was the consumer mantra of the early sixties as miniaturization "spin-offs" found their way into every level of daily life. In the laboratories of the new multinational industries which had grown up around the electronic revolution, the precision of LSI techniques grew through the early 1970s to the point where an entire computer was put on a single chip. In 1971, Intel Corporation introduced its *microprocessor*, based on the design of an employee named M. E. Hoff. By 1975, the third phase of the miniaturization explosion had ignited a secondary conflagration—the microprocessor explosion—when hundreds of ways were found to use these fully operational, cheap, nearly microscopic computer chips.

By 1970, the electronics industry had evolved through several fast cycles of research, development, and diffusion. Because the computer is really a general-purpose tool and only a calculating machine by a trick of history, computers found uses at progressively broader levels of society as miniaturization brought the price of computer power down to manageable levels. In the beginning, only the government could afford to own a computer; a few years later, large corporations, research laboratories, and industries learned what economic prodigies they could accomplish with information appliances. Then schools and middle-level businesses were computerized, followed by small businesses.

By 1973, people were predicting a new market for *personal* computers. In 1975, a tiny company in New Mexico put the Intel 8080 chip together with other necessary components and advertised the first computer kit for $420. They were immediately swamped with orders; other companies soon jumped into the act, and personal computers are now a billion-dollar business. Predicting the impact of such widespread computer power is notoriously difficult. When the price of the automobile reached the range of the average citizen in the era of the

Model T, who could have predicted what personal automobiles would do to our lives? The highway system hadn't been built yet. Where will the microprocessor take us in the decades ahead?

CONVERSATIONAL TACTICS:

With a fast-moving technology like solid-state electronics, there are two essential questions to ask: "What's new?" and "What's next?" In case somebody asks *you* one of these essential questions, just say, "VLSI and integrated optics, probably." While the general citizenry is keeping busy with the abrupt social changes that are following the LSI phase, basic research efforts have gone on to *Very* Large Scale Integration (VLSI), in which computer-directed lasers or particle beams etch up to 250,000 elements on single chips even smaller than LSI circuits; in VLSI, the P and N zones are grown right into the molecular structure of the crystals, and computer techniques allow photomasks to be reduced even further in size.

Beyond VLSI are a multitude of now exotic technologies, many of which will become the household words of the 1990s. Integrated optical systems will use light guides and gates instead of conventional circuitry, will employ laser light instead of electrons, and may be able to store ten thousand times as much information as any electronic system. On another development track are superconducting microcircuits which will take advantage of the special physical effects of extremely low temperature semiconductors to create computers capable of operating at very high speed with very low energy input.

If it weren't for transistor radios, moon landings, heart pacemakers, jumbo jets, satellite-transmitted television, video games, and digital watches, some of the predictions for the coming decades might seem fanciful. If the social upheavals of the past forty years have taught us anything, nobody should doubt that we'll have a new, as yet unnamed, revolution on our hands when the ultraminiature citizens-band mind amplifier and portable library filters from the laboratories and lands in our living rooms.

RELATED TOPICS:

Analog Computer, Artificial Intelligence, Digital Computer, Fiber Optics, Information, Laser.

FURTHER READING:

The Micro Millennium, Christopher Evans, Viking, 1980.
"Teletext and Technology," Daniel Bell, *Encounter*, June, 1977.

MUTATION

DEFINITION:

An event that gives rise to a heritable alteration in the genotype.

WHAT IT REALLY MEANS:

Like "cancer," "mutation" is one of science's bogey words. It is associated with horrible deformities and diseases. As if nature's lottery weren't cruel enough, scientists discovered about fifty-five years ago that X rays can induce mutations. As a consequence, humans now have it in their power to cause a greater number of mutations in a single generation than nature can cause in one thousand generations. Even though mutations are generally harmful to the individual organisms which exhibit them, mutations are also essential to the viability of the species. "Mutation" is a scare word because only half the story has been publicized.

The word itself is derived from the Latin *mutare*, meaning "to change." Variations, or changes, among species of domesticated plants and animals have been acknowledged by farmers and herdsmen for thousands of years. Without a theory of evolution or any understanding of genetic mutation, ancient farmers were able to increase crop yields by sowing the seeds of plumper, mutant cereal grains. Crop surpluses, the historians remind us, were the economic basis of the first civilizations.

Mutations are much more vital than mere producers of a few bushels of corn. The essence of evolution is variation and selection. Mutation produces gene changes in a small number of individuals, and nature's survival requirements endow a few of these changes with greater environmental viability; because they enhance the individual's chance to survive and pass on the characteristic, beneficial genes spread through the species. Even though it is true that most mutations, including the dreaded lethals, are not improvements in the gene pool, it is only through the small number of beneficial mutations that any species improves and becomes better adapted to changing circumstances. Without mutations, all species, our own included, would be far more vulnerable to extinction.

Until very recently (on an evolutionary time scale), all selection was controlled by natural forces. This dominance by nature is now challenged by human technology. According to the philosopher-scientist Jacob Bronowski, the unique thing about us is our ability to shape, rather than simply be shaped by, the environment. This is graphically illustrated by the case of the ancient farmers: the plumper, mutant cereal grains may not have been more viable if left solely to nature, but humans intervened and created an environment in which those plants would thrive. In a more recent, less advantageous case, human use of pesticides has created species of superbugs—mutant insects immune to certain poisons who thrive in the new environment where their previously more numerous, "normal" brethren were killed off. The important question, of course, is: Are we wise enough to be sure our massive environmental alterations are ultimately beneficial?

Human power over the evolutionary process is not confined solely to the selection phase. In 1927, H. J. Muller demonstrated that X rays could induce mutations in fruit flies. We now know that all forms of ionizing radiation are mutagenic. Although Muller's discovery was revolutionary in the science of genetics (it enabled geneticists to produce mutations on demand and facilitated the study of the relationship between genes and mutation and the nature of the mutation process itself), it has also provided one of our most deadly threats: thermonuclear war.

Since the atom was "harnessed" in 1945, the amount of radioactive, mutagenic material in circulation has increased steadily. We know that the rate of occurrence of mutation has a direct relationship with the dose of radiation received and that the effect of these dosages is cumulative. In short, our increased use of radiation sources will hasten the rate of mutation. Toxic chemical mutagens are also infiltrating our environment at an alarming rate. Because most mutations do not improve the gene pool, we are engaging in a biological form of Gresham's Law: Bad genes drive out the good genes. It is difficult to see how this could be beneficial for the long-term survival of our species.

CONVERSATIONAL TACTICS:

In the 1950s and early 1960s, Hollywood seemed convinced that if we didn't blow ourselves up, then we would at least be visited by a succession of mutagenic monsters. Ironically, the Japanese cinema also picked up the killer-mutant theme. Next time your drinking associates start to worry about nuclear Armageddon and attendant scary mutations, challenge them to this movie-monster trivia quiz. You'll need to get at least seven out of ten correct to be considered a true movie-mutation master.

1. Start with an easy one. What type of monster posed the threat in *Them!*?

2. How was The Blob finally killed?

3. Fill in the last two words in the title of this 1957 science-fiction movie classic: The Incredible ________ ____.

4. Match the movie monster with its description:

a. Ghidra	x. Giant jellyfish
b. Dagora	y. Three-headed
c. Gamera	z. Toothed turtle

5. Rodan was a very successful Japanese import. What kind of monster was it?

6. What now-famous actor played the lead role in *I Was a Teenage Werewolf*?

7. Abbott and Costello met a lot of creepy characters in their movies. Name at least three.

8. Everyone remembers Mothra, but do you remember who summoned her in times of need?

9. What famous line is offered by the fly with a man's head in the movie *The Fly*?

10. Finish with another easy one. What technological gimmick caused you to wear funny glasses at the movies?

ANSWERS:

1. Giant ants.
2. A trick question. It was never really killed. Instead, it was dropped

on Antarctica, turning into an ominous question-mark shape in the last frame.

3. *Shrinking Man.*
4. a–y, c–x, b–z.
5. A huge, prehistoric bird.
6. Michael Landon.
7. Dr. Jekyll and Mr. Hyde, Frankenstein, the Invisible Man, the Mummy.
8. Two tiny Asian maidens, singing in high-pitched voices.
9. "Help me!" while trapped in a spider's web.
10. 3-D.

RELATED TOPICS:

Altruism/Selfish Gene, DNA, Electronic Smog, Meltdown, Origin of Life, Sociobiology, Toxic Chemicals.

FURTHER READING:

Genetics, R. P. Levine, Holt, Rinehart and Winston, 1962.
Mutations: Biology and Society, ed., Dwain N. Walcher, Masson, 1978.

NEUTRINOS

DEFINITION:

A subatomic particle of the lepton family with the characteristics of zero mass, one-half spin, neutral charge, stable mode of decay, and an infinite average lifetime.

WHAT IT REALLY MEANS:

Particle physics has produced a swarm of over two hundred subatomic entities, and quantum theory has predicted many phenomena which seem to defy common sense, but the neutrino is perhaps the strangest denizen of either of those scientific realms. Because it is as close to nothing as something can get and still remain "some thing," neutrinos have great theoretical significance. They also promise to have practical applications—as a way of predicting how the universe might end, as a means of measuring events in the core of our sun, and as a potential answer to the problems plaguing nuclear fusion research.

Neutrons are electrically neutral particles, which, along with positively charged protons, form the nucleus of atoms. When a neutron becomes unstable and starts to disintegrate—a process called *beta decay*—it emits a negatively charged electron and is transformed into a positively charged proton. When quantum physicists tried to explain this process, they were deeply troubled by the fact that the decay of the neutron always ended up with some leftover energy and momentum that their equations did not predict. This disagreement between observation and theory threatened the basic physical laws of "conserved quantities" which formed the foundation of all physics. Beta decay was the occasion for a true scientific crisis.

In an admittedly "desperate measure" to explain the excess energy and momentum in beta decay, Wolfgang Pauli proposed in 1930 that another particle—as yet undetected by experiment—must also be emitted when a neutron disintegrates. In order for the equations to balance, the particle would have to be extraordinarily light, electrically neutral, and have a spin of one half. This purely theoretical entity was dubbed *neutrino*—"the little neutral one"—and its actual physical existence was long in doubt. Yet in 1957, Pauli's desperate theoretical measure was transformed into an accurate prediction. American physicists detected the "ghost particle" and triumphantly cabled Pauli the news—twenty-seven years after his prediction.

Once everyone was convinced that neutrinos actually existed, it didn't take scientists very long to put them to work. Because almost everything in the universe is transparent to the passage of neutrinos, they offer an excellent opportunity for observing previously unobservable places like the interiors of stars, distant galactic catastrophes, and events taking place in the core of our sun. Of course, no one gets to observe these phenomena directly; instead, the neutrinos do the observing and "report back" via a neutrino detector. This is especially useful in studying what goes on deep inside our sun, for the densities at the solar core are so great that it takes light millions of years to get to the surface. To the neutrino, however, even such unimaginable densities are almost like empty space.

About fifteen years ago, a scientist named Raymond Davis suggested building a neutrino detector to test theories about the way the sun works. The detector itself turned out to be one of the more bizarre contraptions ever concocted by experimental physicists—a tank full of cleaning fluid, buried deep inside a mine in Colorado. The reason for putting it 1.6 kilometers deep in a gold mine was to shield out all other particles that lacked the neutrinos' penetration power. The cleaning fluid contained a known amount of chlorine-37. Whenever an atom of chlorine-37 is struck by a neutrino, it is transformed to radioactive argon-37. Sophisticated radiation detectors made it possible to count, literally atom by atom, the amount of radioactive argon produced and therefore to estimate the number of neutrinos entering the tank.

In every thermonuclear fusion reaction, four atoms of hydrogen combine to form one helium atom, releasing two neutrinos. By calculating the amount of fusion taking place in the sun, scientists predicted that we should receive, here on earth, around sixty billion neutrinos per square centimeter per second. By multiplying the probability of any one neutrino colliding with a chlorine nucleus by the estimated number of neutrinos radiating from the sun, they estimated the number of argon atoms which should have been detected. When the results were checked, rechecked, and rechecked again, a startling fact emerged: Only about one third as many neutrinos were detected as originally predicted. The consequences could be ominous. Either something has gone wrong with the sun or something is wrong with fusion theory. Or perhaps something is lacking in quantum theory.

In an attempt to explain this new problem—just as Pauli originally predicted the neutrino to explain the beta decay problem—physicists are now proposing that there may be three forms of neutrino. The final results aren't in yet, but if it turns out that one form of neutrino does have a finite but very small mass, it could explain the sun's missing neutrinos. It would also imply that there is enough mass in the universe for the present expansion to end and the universe to eventually contract into another big bang. But if the neutrino is found to have a definite mass, the energy-matter conservation laws would again be in jeopardy.

Neutrinos—Invisible Pollution

Either way, neutrinos will be a hot topic among physicists for years to come.

CONVERSATIONAL TACTICS:

Neutrinos are not so distant as you might think. If you attempt to explain them, and your conversational companion thinks that high-speed bundles of nothingness are a bit too esoteric to be of concern to an ordinary earthling, you might mention that every human is penetrated by billions of neutrinos every second! There is no cause for alarm about "neutrino pollution," though. Because they are so resistant to all the normal barriers, the average neutrino must travel through approximately one hundred million miles of lead before it collides with any other particle.

All subatomic particles, particularly neutrinos, are whimsical objects. In some circumstances, it might be wise to temper explanation with whimsy, as exhibited in this limerick:

The poet, J. Alfred Neutrino,
Who subsisted sublimely on vino,
With a spin of one-half
Wrote his own epitaph:
"No rest-mass, no charge, no bambino."

—Jole Haág

RELATED TOPICS:

Cosmology, Fusion, Quantum Theory, Quarks.

FURTHER READING:

"Missing Particles Cast Doubt on Our Solar Theories," James S. Trefil, *Smithsonian*, March, 1978.
"Ups and Downs of Neutrino Oscillation," Dietrick E. Thomsen, *Science News*, June 14, 1980.

ORIGIN OF LIFE

DEFINITION:

The systematic study of the conditions and processes which led from inorganic matter to organic molecules and, finally, to primitive living organisms.

WHAT IT REALLY MEANS:

Darwin's theory of evolution offers a bold solution to an important mystery—how single-celled life forms managed to grow, over billions of years, into walking, talking, thinking beings. The ideas of variation and natural selection furnished the mechanism by which that miraculous journey from the slime to the threshold of the stars could have been accomplished without supernatural intervention. The theory doesn't say much, however, about an even greater mystery—how life arose in the first place. In order to remedy that deficiency in our knowledge, biochemists, astrophysicists, and geologists are piecing together a scenario covering those crucial years *before* Darwin's grand drama began.

Strangely enough, the first large-scale formal effort to assemble the disparate evidence from various disciplines was initiated by NASA. As America's space agency, one important aspect of its program is the search for extraterrestrial life. It decided that the best way to accomplish this was to understand the events which led to the origins of life on earth. When the *Viking* Mars lander was designed (in part to perform experiments probing the possibility of life on that planet), it became necessary to define "life" in terms that a machine could understand. The working definition of life at its simplest level, official NASA version, is: *a carbon-based macromolecule capable of replicating and metabolizing.*

Replication—the ability to reproduce—is the key criterion for distinguishing a living molecule from an inert molecule. When atoms group together into molecular chains, the resulting compounds can have amazing properties, but one thing a nonliving molecule *cannot* do is make a duplicate of itself. Living molecules *can* reproduce themselves because of the presence of complex molecules called nucleic acids. Since every living creature contains nucleic acids in every cell, the map to the evolution of higher life forms can only be deciphered by understanding how these molecular patterns duplicate themselves.

Life may be viewed as a series of specific chemical messages passed through time, written in a tantalizing code. When the self-replicating mechanism of the DNA molecule was finally explained in the 1950s, that code was finally deciphered. The laboratory search for the origins of life then shifted and now concentrates on how natural processes could have combined simple compounds to form the first DNA.

All life on earth, from the smallest microbes to "intelligent" life, has a common chemical ancestry. This historical continuum of ever more complex, increasingly information-rich, chemical steps can be traced, theoretically, back to the creation of the universe. When our solar system coalesced from huge clouds of interstellar debris approximately four billion years ago, the complex atoms necessary for the later evolution of life were included in that conglomeration because they had been ejected from the cores of exploding stars billions of years previously. As the earth began to cool and condense over the span of a billion years, geological processes began the first steps toward biological evolution. Necessary elements rose to the surface; volcanoes spewed other ingredients into the newborn atmosphere; water began to condense into liquid form.

Our earliest common ancestor, that magical chain of molecules which learned to pass itself beyond death and through time, was born in a primordial soup between three and a half and four billion years ago. In a real sense, that talented compound was a set of immortal instructions, slowly constructed from an alphabet of elements. Carbon, hydrogen, nitrogen, oxygen, and other needed atoms were present in that ancestral broth, to be bombarded by ultraviolet and cosmic radiation, ionized by superbolts of lightning, gently heated by the earth's molten core.

Laboratory experiments have combined these same simple elements, subjected them to similar conditions, and succeeded in demonstrating the natural formation of the building blocks of proteins known as amino acids. It is now believed that further random stimuli fused the amino acids into longer, even more complex molecules that qualified as the first nucleotides. These long chainlike molecules then combined in just the right way to make the first DNA molecule. That was the end of random processes as the fuel for the evolutionary engine; the first DNA molecule used the nucleotides in the soup surrounding it to make the second DNA molecule . . . and biology was off and running.

In addition to their laboratory simulations of primordial conditions, scientists have also been searching for fossil evidence of the earliest life forms. Ironically, the two oldest and most exciting finds in the search for life's origin have been made in two of the most lifeless areas of the earth—a place in Australia called "the North Pole" because of its remoteness, and a rock formation in Greenland known by the Eskimo name *Isua*, meaning "the farthest one can go." The fossils in Australia, technically known as *stromatolites*, were discovered in 1980, and they clearly indicate the presence of bacteria at least three and a half billion years ago. Radioisotope dating of the Greenland rocks indicate that hydrocarbons likely to have been created by living organisms were present three hundred million years before the Australia fossils.

CONVERSATIONAL TACTICS:

Genesis says that creation was accomplished in seven days. Scientists sometimes pretend to take this schedule literally in order to demonstrate

Origins of Life—
Life's Calendar

the relative time scale of the events leading to our present existence. A calendar, an appointment book, or a piece of paper marked off into seven "cosmic days" can furnish an ideal conversational visual aid for discussing the biochemical version of creation.

Begin by explaining that you are going to compress four and a half billion years into one week. By that timetable, the earth and the solar system coalesced sometime very early Sunday morning. In the predawn hours, the planet began to cool, and volcanoes vented gases. Sometime between dawn and dusk on Sunday, a primitive atmosphere and ocean had formed. In the waning hours of the first night, lightning, solar radiation, cosmic radiation, and geothermal energy sparked the creation of prebiotic molecules in a slightly acidic soup.

Two A.M. Monday corresponds to three billion, eight hundred million years ago, when microorganisms might have been present in the Greenland rocks. The Australian bacterial colonies spread across the ocean floor at one P.M. Monday. The earliest remains of photosynthesizing bacteria were laid down sometime Tuesday afternoon. Around nine P.M. Wednesday (two billion years ago), fossil evidence found in what is now Gunflint, Canada, indicates an explosive diversity of bacteria.

At eight P.M. Thursday, a momentous event occurred—the first cells with nuclei appeared. With the exception of bacteria and blue-green algae, every plant and animal since then evolved from those nucleated cells. Most of the action took place during the weekend: for one thing, photosynthesis in marine plants created an oxygen atmosphere. At one thirty A.M. Saturday, marine life flourished, and trilobites ruled the seas. By nine o'clock Saturday morning, jawed fish and small land plants appeared. Sometime in the morning hours of the seventh day of creation, something crawled out onto dry land and learned how to breathe—or, more likely, the seas receded and left millions of candidates with the choice of evolution or extinction.

Two hours and two hundred million years later, amphibians marked the large-scale transition to dry land. Swamps, forests, and insects covered large areas of the earth. At five P.M. Saturday afternoon, the reptiles and dinosaurs appeared, expanded, and ruled. When they died out suddenly and mysteriously, after hundreds of millions of years of dominion, mammals evolved to take the top rung on the terrestrial ladder.

At this point in your story, draw a clock face in the corner of the seventh calendar page. With only minutes left in the week, the period most specifically covered by Darwin happened: At about ten fifty-three

P.M., thirty million years ago, the oldest common ancestor of apes and men emerged in the forests which then covered Africa; *Homo erectus* walked out of the trees and onto the savanna at three minutes before midnight; and *Homo sapiens* began to build civilizations four seconds before midnight. All of recorded history has to be compressed into a fraction of the last second of the week.

RELATED TOPICS:

Altruism/Selfish Gene, DNA, Entropy, Exobiology, Hot-Blooded Dinosaurs, Human Origins, Photosynthesis, Sociobiology, Viruses.

FURTHER READING:

The Roots of Life, Mahlon B. Hoagland, Avon, 1979.
"In The Beginning," Joel Gurin, *Science 80*, July/August, 1980.

PARTICLE ACCELERATOR

DEFINITION:

One of a variety of electrical devices which impart large kinetic energies to charged subatomic particles.

WHAT IT REALLY MEANS:

Advances in particle physics rely on a kind of bootstrap operation between experimenters and theoreticians. Sometimes an experiment reveals an unsuspected property of matter; it is then the job of the theoretician to explain how the observed phenomenon fits in with other known properties. At other times, a theory is postulated which predicts an as yet undetected feature of matter at the atomic level; when this occurs, it is the job of the experimenter to try to find evidence of such particles.

Just such a bootstrap operation is responsible for the discovery of the neutrino. Experimentalists first noticed that when neutrons decayed, a small amount of mass-energy was left unaccounted for. Theoretician Wolfgang Pauli tried to explain this anomaly by hypothesizing the existence of a new particle, the neutrino. More than twenty-five years later, the existence of neutrinos was verified by careful experimentation. While theoreticians work with chalk and blackboards, the particle accelerator, or "atom smasher," is the chief tool of the experimentalist.

The method used by particle physicists has been called the "Swiss watch method": If you want to find out what is inside two watches but are prohibited from taking them apart and looking inside, you simply bang them together and see what comes out. Particle physicists are interested in the insides of atoms and need a way of banging them together. The accelerator is just such a way. Using tremendously high energies, enough to strain the power capacity of a decent-sized town, these accelerators speed particles into collisions with atomic nuclei. The impact releases charged particles, which then leave tracks in a bubble chamber. The track is evidence of the existence of a subatomic particle.

There are many different types of accelerators now in operation, and they all seem to have science fiction names: bevatron, cyclotron, proton-synchrotron, and SLAC (Stanford Linear Accelerator). The part that isn't science fiction is the cost and size of these machines. The proton-synchrotron at the Fermilab in Illinois can operate at 500 GeV—that's five hundred *billion* volts! SLAC is two miles long (you drive directly over it on the highway between San Francisco and San Jose), and cost $114 million to build back in 1966, when building high-tech devices was considerably cheaper than it is now. When people talk about Big Science—the notion that science can no longer be carried out by a single, dedicated scientist working alone in a basement laboratory—more often than not they have particle accelerators in mind.

CONVERSATIONAL TACTICS:

Even in this age of Big Science, it is still possible for you to see what goes on in a particle detector for only a few dollars. Of course, you won't be accelerating any particles, but you will be viewing the actual tracks left by subatomic particles which result from radioactive decay. The diffusion cloud chamber (the granddaddy of the modern bubble cham-

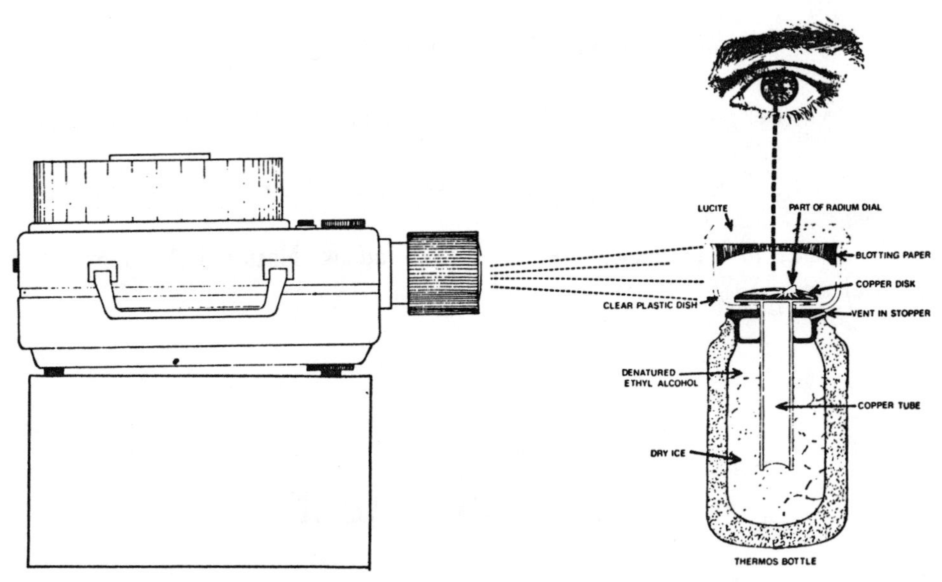

Particle Accelerator—A Homemade Cloud Chamber

ber) works by providing the proper temperature and pressure required for the condensation of vapor in the path of an ionized particle.

Legend has it that the physicist C.T.R. Wilson was inspired to invent the cloud chamber while admiring real clouds, which form as moist air rises and condenses at lower pressure. Forty years later, in 1951, Don Glaser had a similar inspiration over a glass of beer and invented the bubble chamber—a device which employs gas bubbles in a liquid instead of liquid droplets in gas.

The only ingredient for your experiment which isn't readily available is the radioactive source. Radium watch dials, fortunately for the general health, are not as prevalent as they used to be, but you might find one in a secondhand store or pawnshop. While you are shopping, you might also look for orange Fiestaware—a popular dish set from the 1950s. It's no longer popular because it turned out that the orange paint was slightly radioactive! As it happens, though, some manufacturers continue to use weak radiation sources in household objects. There is a very weak source of radioactivity in some kinds of home smoke detectors.

Just follow the diagram, turn out the lights, look closely, and play particle physicist.

RELATED TOPICS:

Fission, Neutrino, Periodic Table, Quarks.

FURTHER READING:

"Splintering the Atom," William Kaufmann III, *Science Digest*, March, 1981.

From X-Rays to Quarks: Modern Physicists and Their Discoveries, Emilio Segre, W. H. Freeman, 1981.

Science and Technology: A Five Year Outlook, National Academy of Sciences, 1979.

PERIODIC TABLE

DEFINITION:

A tabular arrangement of the chemical elements, illustrating that the properties of the elements recur at regular intervals and as a function of their atomic number.

WHAT IT REALLY MEANS:

If you ever took a science class in school, you undoubtedly spent many hours staring at that strange calendarlike chart behind the teacher's desk—the periodic table of the elements. You might even recall that the chart is some kind of key to chemistry, if you could only remember how to read it.

What your teacher or text probably *didn't* tell you is that the chart was invented by a chemistry teacher in nineteenth-century Russia. Dmitri Mendeleev began looking for a way to illustrate the periodic properties of the elements to his students and ended up with an individual perception of cosmic patterns—a vision of order so lucid and unifying that it made a true science of chemistry possible.

What Mendeleev saw that nobody before him saw was a pattern that neatly assembled the chaotic fragments of experimental chemistry into a single theoretical model, somewhat like fitting the individual words into a crossword puzzle. The bold model he constructed had a power even greater than the ability to make sense out of conflicting data. The periodic table made specific predictions about the qualities of unknown elements, which experimentalists could search for in their laboratories. The technology of modern chemistry would not exist today if it were not for that dynamic interplay between Mendeleev's model and the experimental data from the chemistry laboratories of the late nineteenth and early twentieth centuries.

The periodic table is like one of those Tibetan mandalas that bestows enlightenment only if you know the proper way to look at it. This specific way of arranging the elements is the crucial link between medieval alchemy and modern chemistry. In 1869, at the time Mendeleev was writing his text, about sixty elements had been identified and their properties studied. However, no one in the world knew how to make sense of this "elemental" hodgepodge. How many elements could there be? Why did some of them act in similar ways? Was there a "natural order" to this mélange?

It occurred to Mendeleev that if he arranged the elements according to atomic weight and in just the right pattern, then obvious recurrences of similar chemical properties reappeared at regular intervals. He decided to play solitaire with the elements, so he wrote the names and numbers of the known elements on cards and then tried to arrange the cards in an explanatory matrix—the vertical rows could contain elements with similar properties. For example, the first five elements, reading down the left column, are the alkali metals—lithium, sodium, potassium, rubidium, and cesium. The column under helium, at the far right side of the chart, lists neon, argon, krypton, xenon, and radon—the "noble" gases, those that will not combine with other elements. Whereas the alkali metals are highly reactive, the noble gases are known for being inert.

Mendeleev's chart was such an accurate model that it could be used to predict the future discovery of elements which were then just blank spaces in the solitaire design. For example, there was a blank space above aluminum in the chart, suggesting the existence of an undiscovered aluminumlike metal possessing a specific atomic weight. In 1875, a French chemist discovered gallium, but it was too light to fit into the pattern. Mendeleev had the nerve and confidence to ask the French chemist to recheck his measurements. It turned out that the prediction based on the Russian's pattern was correct; the original experiment was slightly inaccurate. Other holes in the table continued to be filled in until, by the end of the nineteenth century, the first

1 H																	2 He
3 Li	4 Be											5 B	6 C	7 N	8 O	9 F	10 Ne
11 Na	12 Mg											13 Al	14 Si	15 P	16 S	17 Cl	18 Ar
19 K	20 Ca	21 Sc	22 Ti	23 V	24 Cr	25 Mn	26 Fe	27 Co	28 Ni	29 Cu	30 Zn	31 Ga	32 Ge	33 As	34 Se	35 Br	36 Kr
37 Rb	38 Sr	39 Y	40 Zr	41 Nb	42 Mo	43 Tc	44 Ru	45 Rh	46 Pd	47 Ag	48 Cd	49 In	50 Sn	51 Sb	52 Te	53 I	54 Xe
55 Cs	56 Ba	57 La	72 Hf	73 Ta	74 W	75 Re	76 Os	77 Ir	78 Pt	79 Au	80 Hg	81 Tl	82 Pb	83 Bi	84 Po	85 At	86 Rn
87 Fr	88 Ra	89 Ac	104 Rf	105 Ha	106	(107)	(108)	(109)	(110)	(111)	(112)	(113)	(114)	(115)	(116)	(117)	(118)
(119)	(120)	(121)	(154)	(155)	(156)	(157)	(158)	(159)	(160)	(161)	(162)	(163)	(164)	(165)	(166)	(167)	(168)

lanthanides	58 Ce	59 Pr	60 Nd	61 Pm	62 Sm	63 Eu	64 Gd	65 Tb	66 Dy	67 Ho	68 Er	69 Tm	70 Yb	71 Lu
actinides	90 Th	91 Pa	92 U	93 Np	94 Pu	95 Am	96 Cm	97 Bk	98 Cf	99 Es	100 Fm	101 Md	102 No	103 Lr
superactinides	(122)	(123)	(124)											(153)

Periodic Table—Mendeleev Completed

radioactive elements were discovered and chemistry branched off into nuclear physics.

The most vital period in the history of chemistry was triggered by the theoretical-experimental framework imposed by the periodic chart. The periodicity of chemical properties and, later, the strange new properties of radioactive elements, led to a deeper level of theory explaining how the elements themselves are constructed. In the first decades of the twentieth century, Rutherford and Bohr established that every atom has a small, dense nucleus, which carries a positive charge (proton), and that the nucleus is surrounded by extremely light, orbiting electrons. Hydrogen, the first element in the table, has one proton and one electron. Helium has two of each, and so on. All the elements heavier than hydrogen also contain neutrons, which have equal mass to the proton but carry no charge. The electrons are so light in comparison to the nucleus that the atom's weight is considered to be a combination of the protons and neutrons.

In 1914, a British physicist named H.G.J. Moseley showed that the periodic sequence of elements is ordered more precisely by just counting the number of *protons* in each nucleus—the *atomic number*. The physical principle underlying the periodic recurrence of properties was then explainable by a model of the atom in which electrons form orbital shells around the nuclei of the elements. A new shell starts after the inner shell is filled, so that in only the outer shell are the electrons (or lack of electrons) able to interact with the extra electrons (or deficiencies) in the outer shells of other atoms. This perfectly explains the cyclic properties: the noble gases, for example, are almost totally inert because their outer shells are filled.

The blank spaces in the periodic table continued to be filled in, until there seemed to be no new elements beyond the ninety-second—uranium. Then, in 1938, Otto Hahn in Germany bombarded uranium with neutrons in order to create a heavier element. When the Austrian nuclear physicist Lise Meitner realized that they had not created a heavier element but instead had *split* the uranium nucleus into two lighter atoms, the process of nuclear fission was discovered, changing the course of human affairs as radically as it changed the old periodic table.

CONVERSATIONAL TACTICS:

The goal of the ancient alchemists—the transmutation of elements—was finally achieved by the nuclear physicists of our century. In the subspecialty of nuclear physics concerned with the creation of transuranium elements, Glenn Seaborg must be regarded as the foremost living alchemist. Of the fourteen elements thus far created at the upper boundaries of the periodic table, Seaborg and his team have been responsible for ten, including that notorious radioactive metal, plutonium. Because the newest elements are highly radioactive, they tend to decay quickly into lighter elements—the most exotic new elements have half-lives of minutes or seconds, or even millionths of seconds. It isn't easy to determine the properties of something which exists for such a short span of time.

As more powerful particle accelerators are built, opening the possibility of new experimental transmutation techniques, the theorists have been inspired to talk about a heretofore hidden section of Mendeleev's old arrangement—an "island of stability" around elements number 110 to 114. If you take the time to explain the periodic table and hear the criticism that it sounds like old news, mention the possibility of a new precious metal in the blank space below the line containing gold and silver. Is there a radioactive noble gas? With the development of subnuclear alchemy, that century-old model of "what the world is made of" may still have some surprises in store for us.

RELATED TOPICS:

Acid Rain, Quantum Theory, Quarks.

FURTHER READING:

The Search for Solutions, Horace Freeland Judson, Holt, Rinehart and Winston, 1980.

The Development of Modern Chemistry, Aaron J. Ihde, Harper & Row, 1964.

PHEROMONES

DEFINITION:

Pheromone, a combination of the Greek roots meaning "to transfer" and "to excite," designates a chemical emitted by one organism which alters the behavior of other organisms of the same species.

WHAT IT REALLY MEANS:

Scientists now think that the *nose* may be man's and woman's most important sex organ. The discovery of pheromones—odors which transmit messages—could herald the most important communication breakthrough since the first cave dweller learned to get what he wanted by grunting rhythmically. It now appears that the nose, along with certain tufts of body hair, may be the antenna for a biospheric communication network, a chemical switchboard linking us to one another and to every other life form on this planet.

Molecular message packets wafting in the breeze may be telling us all what to do, when to do it, and who to do it with. In many lower species, pheromonal regulation of behavior has been rigorously demonstrated. Identity, maturity, territoriality, aggression, homing, nurturing, and sexual bonding are demonstrably pheromone-mediated in scores of species. In humans, the facts about pheromones are only beginning to be revealed.

Until recently, the hows and whys of human pheromones were just conjecture. It is a brand-new science, and the early findings were as intriguing as they were fragmentary: In 1971, Martha McClintock at Harvard observed that coeds in college dormitories synchronized their menstrual cycles; a correlation has long been known to relate female estrogen cycles and sensitivity to certain odors; in West Germany, an aerosol aphrodisiac for pigs was marketed. The pig attractant, now routinely used to prepare sows for artificial insemination, was known chemically as *androstenone*.

In the late 1970s, androstenone-soaked handkerchiefs were sold in British sex shops under the label Aeolus-7. In mid-1980, a much more sophisticated variant appeared on the shelves—an odorless spray marketed under the name Bodywise. The key ingredient in this putative human aphrodisiac is a relative of androstenone, a substance named *alpha androstenol*. It is also related to the androgens, the male sex hormones.

In 1979, a team at Warwick University, led by George Dodd, identified two compounds in human sweat. They managed to synthesize one of the compounds—alpha androstenol—and it has been tested in a number of ways. When it was clandestinely sprayed in the air, both men and women rated photographs of women to be sexier and more attractive than the same photographs which were judged in nonsprayed rooms. When it was sprayed on theater seats, chairs in waiting rooms, and telephone booths, there was a marked increase in the time of occupancy by both sexes, but women appeared to be more strongly drawn to those places.

There is an overwhelming intuitive reason for investigating human pheromones. What else could account for the volative and mysterious "chemistry" which happens between people? Sex isn't the only important human activity that might be linked to olfactory alchemy. Rage, mood, aggressive behavior, and fear are also involved in the same ancient subsection of the human nervous system known as the "smell brain."

Insect experts are notoriously touchy about human analogies, particularly where "instinctive" factors like pheromones are concerned. But the speculative possibilities are too fascinating to ignore. Take ants, for example. Ants as isolated individuals have been likened to nerve cells on legs. As a group, though, ants are far too human for comfort. They practice agriculture, raise livestock, wage war, employ slave labor, bridge rivers, and build high-rise apartments.

Termites are particularly "intelligent." Watch a line of termites and you'll notice that they frequently rub antennae. When two or three randomly groping termites stumble into proximity, an exchange of pheromones directs them into a slightly more ordered unit. As more workers drift into the project, a kind of community odor accumulates, triggering progressively more complex chains of behavior. Mindless, individual food-seeking becomes an orderly supply line. A bunch of insects pushing around grains of sand slowly becomes a team of master architects.

The mind which directs this process is not easy to imagine, for the termite brain does not reside in the nerve tissue of the individual termites but in a network of pheromones. The mind of a termite mound is a silent, invisible, chemical telegraph system; one termite can only transmit one letter, but three can send a syllable, and a million can formulate military strategy.

Pheromone-regulated behavior is not confined to the insect kingdom. Spawning salmon are able to recognize the taste of the old home stream in dilutions weaker than one part per billion. The next time you are served smoked salmon, don't forget that your hors d'oeuvre had a homing mechanism more intelligent than a cruise missile. Once they cross a chemical boundary of another pack, wolf packs will abandon pursuit of wounded game, even in the midst of a famine. Dogs urinate on your leg because you are a territorial boundary marker, while fire hydrants are canine olfactory summit conferences.

Pheromones seem to act as a communal adhesive, holding together and shaping the group behavior of packs, swarms, schools, and flocks of insects and animals. Could they function similarly in families, clans, tribes, and nations of humans? If chemical regulation turns out to be one tenth as pervasive in our own species as it is in so many others, then we humans may have to recast our fundamental beliefs about sex, free will, society, and the future of civilization. The pheromonal regulation of behavior may be the crucial test for many of the more radical sociobiological theories.

Although the science of pheromones is still struggling to be born, a billion-dollar industry is devoted to the promotion of products intended to mask or destroy human odors. The most alarming side effect of deodorant pollution is the evidence that human sex-linked odors are produced with the cooperation of symbiotic bacteria—those nasty little "underarm germs" we are still being trained to kill. Our conveniently moist skin folds and strategically located patches of hair are at least circumstantial evidence that these bacteria are true symbiotic partners who may, in fact, be absolutely necessary for the human sex act. When our underarm symbiotes are eliminated, will we still remember how to reproduce?

CONVERSATIONAL TACTICS:

If you don't have a vial of alpha androstenol, the effect of your own pheromones on the opposite sex can be tested with nothing more elaborate than a small handkerchief and a lot of nerve. Casanova himself discovered this trick long before the discovery of pheromones, and we all know how well he did. Krafft-Ebing, that brilliant, bizarre, eroto-scientific commentator of the nineteenth century, included the "stinky hanky routine" in his monumental compendium of aberration, *Psychopathia Sexualis*: "A sensual young peasant revealed that he had excited many a chaste girl sexually, and easily gained his end, by carrying a handkerchief in his axilla for a time, while dancing, and then wiping his partner's perspiring face with it."

The *axilla* is a medical euphemism for the armpit, so the venturesome experimenter runs the risk of offending the object of the maneuver. Remember, though, that Casanova never got through a successful night without a bold gamble or two.

RELATED TOPICS:

Behavior Modification, Bioethics, Endorphins, Reproductive Technology, Sociobiology.

FURTHER READING:

The Lives of a Cell, Lewis Thomas, Bantam, 1974.

"The Likelihood of Human Pheromones," Alex Comfort, *Nature*, 229:244–45, 1971.

"How to Win the Mating Game—By a Nose," Jo Durden-Smith, *Next*, November-December, 1980.

PHOTOSYNTHESIS

DEFINITION:

The process by which green plants manufacture carbohydrates from carbon dioxide and water, utilizing the energy of light.

WHAT IT REALLY MEANS:

An old joke asserts that there are two kinds of people in the world—those who classify people into two kinds, and those who don't. It may not be too funny as a joke, but it's a good introduction to the arcane world of the taxonomist, that breed of biologist who attempts to classify all of nature's organisms according to kingdoms and phyla and the like. Artistotle was the first great natural scientist, and he developed a taxonomic distinction which has lasted for two thousand years: There are two kingdoms, two forms of life—the plants and the animals.

Unfortunately, Aristotle was not precise enough. Even though we can almost always tell a tree from the toad sitting in it, modern biologists have discovered classes of unicellular organisms which seem to be both plant and animal, or neither. This has sent them scurrying for a new distinction, a kind of cellular difference between plants and animals. The new taxonomic law is this: The energy-extracting mechanisms of living cells are of two kinds, *heterotrophic* and *autotrophic*.

Heterotrophic cells include those of the human body and higher animals in general. They require a supply of preformed, ready-made fuel such as carbohydrates, proteins, and fat. Heterotrophic cells obtain their energy by burning or oxidizing these fuels through the processes of respiration and metabolism, which require atmospheric oxygen. Heterotrophic organisms use this energy to carry out their biological work, and excrete carbon dioxide to the atmosphere as a waste product.

Autotrophic, or "self-reliant," cells obtain their energy directly from sunlight. The energy is used to incorporate carbon from atmospheric carbon dioxide in order to produce glucose, an elementary sugar. By using the glucose as a form of chemical energy, the cells of green plants and other autotrophic organisms are able to build up the more complex molecules of which cells are made. Oxygen is ultimately discharged as a waste product of these organisms.

It should be clear from the above description that *all* living things derive their energy from sunlight. Plant cells can do this directly, animals only by eating the plants or other animals. The reason why plants are autotrophic is the elegant reaction known as *photosynthesis*—literally, "light fusion." Like many important scientific discoveries, the phenomenon of photosynthesis was known and studied long before it was explained. In 1772, Joseph Priestley, the discoverer of oxygen, reported the results of an experiment by concluding that "instead of affecting the air in the same manner with animal respiration, [plants] reverse the effect of breathing." However, it was only through the modern techniques of cellular and molecular biology that we have been able to uncover the step-by-step process of photosynthesis.

That process is actually two related processes—the light and the dark reactions. The light reaction, so named since it occurs when the plant is actually absorbing sunlight, is a photochemical process that takes the sun's light energy and transforms it into stored chemical energy. This is accomplished through another process known as *photophosphorylation*. Plants are green because they contain *chloroplasts*, cellular bodies which contain molecules of *chlorophyll*. Because most of the usable visible light is violet and blue, chlorophyll absorbs those rays while reflecting the blue-green and green wavelengths; hence the green color of plants. Chlorophyll's essential property is the ability to absorb light and use that light to excite electrons within its own atomic structure. These excited electrons (a form of light energy because electrons can jump energy levels only by absorbing or emitting photons) are then shunted through a complicated pathway of reactions which results in the production of adenosine triphosphates (ATPs). ATPs are the major form of chemical energy currency for all living organisms—literally, the elixir of life.

The dark phase, which naturally does not require strong light, uses the stored energy in the ATPs to break the strong chemical bonds between elements while forming weaker bonds between different arrangements of the same elements. The dark cycle also fixes carbon dioxide (CO_2).

Although the exact description of the dark phase is enormously long and complicated, involving some twenty-five distinct reactions and as many intermediate compounds, the gist of the general photosynthetic reaction is quite simple. Carbon dioxide plus water yields glucose (a simple sugar) and oxygen. This simple formula explains how we all eat and breathe.

CONVERSATIONAL TACTICS:

Photosynthesis is not your common, garden-variety conversational topic. Although it affects all of us, it's not sexy, or the cause of corruption in high places, or as familiar as the weather. Nevertheless, with energy and food production becoming topics of growing interest, it shouldn't be long before photosynthesis becomes a story on the six o'clock news. Here's the information you'll need to scoop the newscaster.

Since plants are our ultimate source of food, and since plants produce food (sugars, carbohydrates) through photosynthesis, you don't need to be a Nobel prize winner to see that if photosynthetic efficiency can be increased, so can food production. The simplest approach to doing this is to alter the plants' external structure. When plant strains with a more upright alignment of leaves are selected, more light can penetrate to all the leaves and more ATP can be made. This approach has already yielded a number of hybrid maize (corn) varieties, and work is continuing on other plant varieties.

A second approach is to breed plants which can convert more of their share of sunlight into chemical energy. Currently, the most efficient crop plants operate at 5 percent, while some algae have a 25 percent efficiency rate. Although the light reaction is unusually efficient when compared with most other photochemical reactions, there is some evidence that it could be technologically supercharged. Finally, by manipulating the enzymes during the dark reaction, there is the possibility of improving the plant's utilization of energy once it has converted and stored it. If any of these schemes succeed, we'll have to face the "What have we done?" dilemma. Previous attempts at messing with mother nature have had unexpected, sometimes disastrous side effects.

On the energy front, you can always interject the fact that plants are nature's form of solar batteries—devices which can store light energy as chemical energy. While the physicists and chemists are trying to figure out photovoltaic cells, the biologists and photochemists are trying to mimic the photosynthetic process in laboratory glassware instead of the interior of a plant cell. If artificial chloroplast systems can be constructed on a grand enough scale, the energy crisis will be over. Work has already begun on these synthetic chlorophyll substitutes, with some surprisingly good early results. According to Michael Gräetzel, a leading researcher, "there is a lot of credit, a lot of money, and several Nobel prizes at stake. When water is finally photochemically and efficiently split, the world's energy problems will be solved." If and when that happens, it is safe to assume that we'll all hear about it on the news.

RELATED TOPICS:

Acid Rain, Energy Alternatives, Entropy, Origin of Life, Sunspots, Weather Modification.

FURTHER READING:

Plants at Work, F. C. Steward, Addison-Wesley, 1964.

"The Evolution of the Earliest Cells," J. William Schopf, *Scientific American*, September, 1978.

PLACEBO

DEFINITION:

The word is derived from the Latin verb meaning "I shall please"; a placebo is a physiologically inactive medication prescribed to placate a patient or create a control situation in a drug experiment.

WHAT IT REALLY MEANS:

A placebo is a sneaky way your doctor treats you by getting your body and mind to do their own healing. Since the word *placebo* is connected to emotional effects—*pleasing* rather than curing—it describes something much more complex than a fake pill. Placebos imply entire belief systems. As such, the placebo effect is really a process, a healing ritual involving well-meaning deception. It may also become a major route to the medicine of the future.

The placebo seems to represent a natural phenomenon doctors and shamans have secretly known about for centuries—that physician inside each person who knows how to lower blood pressure, dissolve kidney stones, shrink tumors, or relieve migraine headaches. The problem most of us have is that we simply don't know how to get in touch with our inner medical center. What we need is a link between the will to live and the machinery of the body. As the symbolic messenger of curative power, the placebo pill can supply that link.

Self-healing has long remained in the realm of the miraculous rather than the world of science because the physical mechanisms of mind over matter were not amenable to scientific investigation—not until recently, that is. A new breakthrough in brain biochemistry, the discovery of peptide neurohormones such as the endorphins, strongly suggests that the placebo effect has a measurable organic component. In experiments at the University of California, San Francisco, investigators first administered placebos to patients who complained of pain after wisdom-tooth extractions. As expected from previous experiments, about one third of the placebo-treated group experienced some relief—even though their medication was totally inactive. These patients were then administered a drug that blocks the action of beta-endorphin, a chemical in the brain which is stimulated by painkilling medications. Their pain then returned to expected levels.

The implications of the UC experiment are startling: Placebos might work because they trick patients into producing their own painkillers. By causing the mind to believe it has received a pharmacological painkiller, the placebo activates the brain to produce its own medicine. Further research is under way to find the exact metabolic pathways by which mental suggestion is translated into biochemical self-manipulation.

CONVERSATIONAL TACTICS:

Topics like the placebo effect never fail to produce their share of skeptics. Should you meet one, do *not* try to prove your case by administering your own placebo. Since the chances of success are believed to be proportionate to the quality of a patient's relationship with his or her doctor, it's likely you'll fail, and the skeptic will be smugly justified. Instead, rely on reasoned argument—because all the experimental evidence is in your favor.

The classic study of the placebo effect was conducted by Dr. Henry Beecher at Harvard. He reviewed fifteen separate placebo experiments and found that 35 percent—over one third of the patients studied—experienced "satisfactory relief" from a range of complaints including severe postoperative wound pain and acute seasickness. All that from a sugar pill! Other researchers have found that placebos work for a long list of maladies: arthritis, abnormal blood-cell count, peptic ulcers, hay fever, hypertension, and abnormal respiration. Dr. H. Gold reported, at a 1946 Cornell conference, that people of high intelligence are more likely than others to benefit from a placebo. Try throwing that line back at the skeptics who swear it would never work for them!

One fascinating sidelight to placebo theory is the "mind over warts" phenomenon. If anything in medicine is clearly more physical than mental, it has to be the reproductive apparatus of a virus, which is what a wart really is. However, famed author and immunologist Lewis Thomas reports that studies at Massachusetts General Hospital found that people can be cured of warts through hypnotic suggestion. Science hasn't yet tracked the steps by which a spoken word is transformed into a work order for the immune system, but wart hypnosis is another kind of proof that placebos have a measurable physiological effect.

If you simply want to limit your placebo conversation to a single witty comeback, you might offer this bon mot, attributed to the great nineteenth-century physician Sir William Osler: "The human species is distinguished from the lower orders by its desire to take medicine."

RELATED TOPICS:

Acupuncture, Behavior Modification, Bioethics, Consciousness, Endorphins.

FURTHER READING:

"The Mysterious Placebo: How Mind Helps Medicine Work," Norman Cousins, *Saturday Review,* October 1, 1977.

The Medusa and the Snail, Lewis Thomas, Viking, 1979.

PLATE TECTONICS

DEFINITION:

A unifying geophysical theory which explains phenomena as diverse as deep-sea trenches, earthquakes, volcanoes, and mountain-building. The theory has two components: *tectonics,* the geological processes pertaining to structural deformations in the earth's crust; and *plates,* the gigantic, irregularly shaped, rigid blocks comprising the earth's outer shell.

WHAT IT REALLY MEANS:

The earth is a living, moving being. In the normal geological course of events, mountains are born and die in hundreds of millions of years. Continents drift to even slower tempos. Once in a while, though, one part of terra firma moves just a tiny bit more vigorously than usual, and the terrified survivors witness the most spectacular natural occurrences our planet has to offer—earthquakes, volcanoes, tidal waves.

Volcanic action and earthquakes are the most visible manifestations of plate tectonic action. Despite our belief in "solid earth," most of the inside of our globe consists of a hot, dense, liquid rock layer surrounding a denser and hotter liquid metal core. The outermost rigid crust we walk on is relatively thin, and not at all uniform. The continents themselves rest on enormous plates, each one about two hundred miles thick, which float on the liquid mantle. When two plates rub together, the resulting friction deep underground may create channels for magma to erupt through the surface, causing volcanoes. When the pressure of plates moving in opposition causes parts of the outermost crust to slip suddenly, earthquakes are created.

The scientific field of plate tectonics is still in its infancy. So far, the theory is doing what a good theory should do—it explains, simply and credibly, why and how observable geological events occur. The next step—actual prediction of those events—is awaited anxiously by those people who live near fault zones or active volcanoes.

CONVERSATIONAL TACTICS:

If seeing is truly believing, it should be easy to demonstrate the logic of plate tectonics to any doubter. Simply point out the outlines of Africa and South America on a map or globe and show how they neatly fit together, like pieces of a jigsaw puzzle. This is one confirmation of plate tectonics—and delayed support for the older theory of continental drift. German geologist Alfred Wegener asserted early in this century that all the continents were long ago jammed together into a single land mass, which later separated into the present continents. Wegener was initially ridiculed because almost all of his colleagues thought of the earth as solid and stable, with no internal mechanics which might cause continents to drift. Plate tectonics provided that mechanism, and Wegener is now regarded as a prophet.

The Mesozoic-era supercontinent was given a name—*Pangea.* When Pangea broke into two land masses, there was nobody around to name them, so geologists did it a few millions of years after the fact—*Laurasia* was the name selected for the northern mass, and *Gondwanaland* was the southern mass.

The continents haven't stopped drifting. If plate tectonics holds true in the future as it did in the past, Baja California should be an island off the coast of British Columbia in another fifty million years!

RELATED TOPICS:

Hot-Blooded Dinosaurs, Human Origins, Origins of Life.

FURTHER READING:

"Exploring the Earth Within," Roger Bingham, *Science 80,* September-October, 1980.

Continental Tectonics, National Academy of Science, 1980.

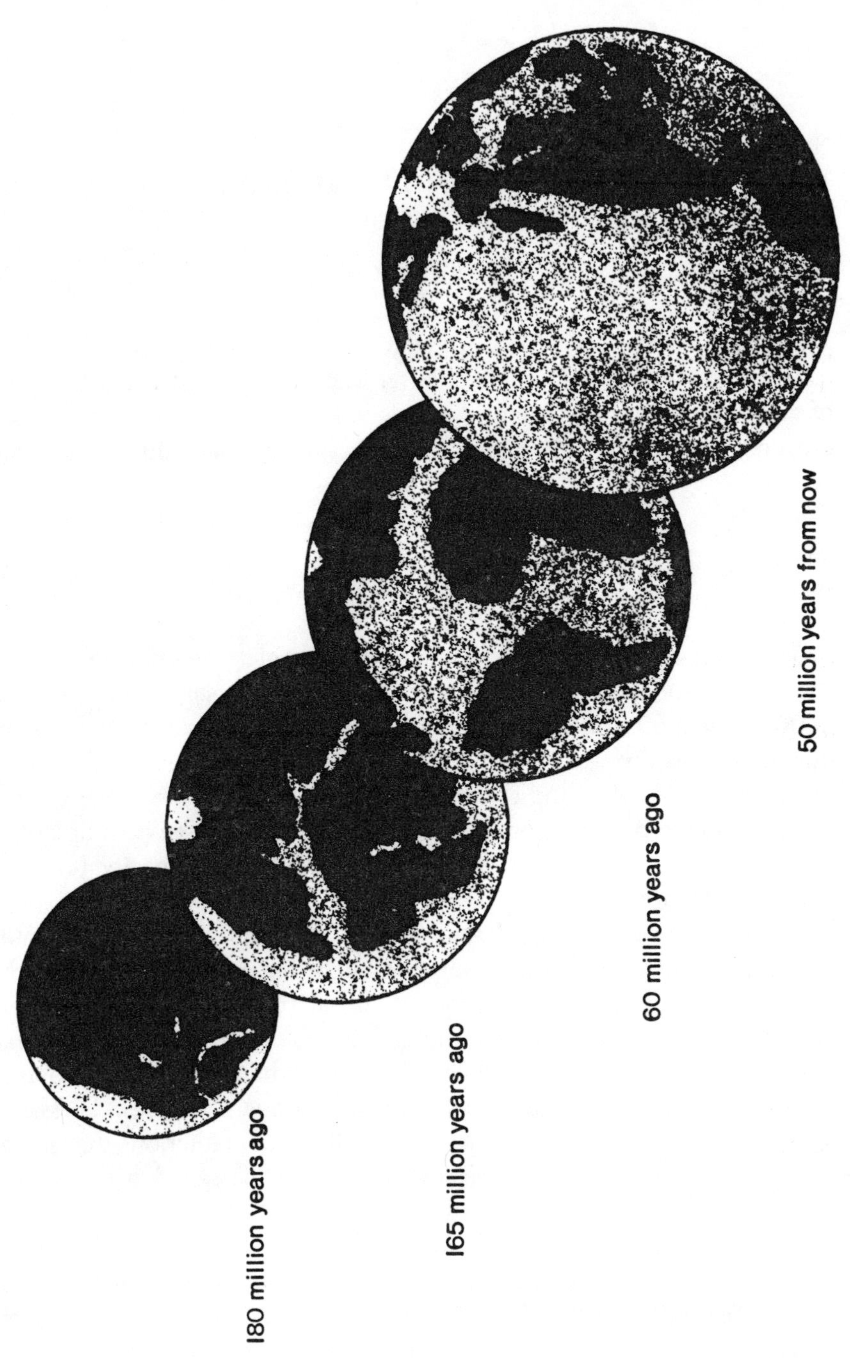

Plate Tectonics—Earth in Transition

QUANTUM THEORY

DEFINITION:

A group of theories which originated from the hypotheses of energy quantization and wave-particle duality. Mathematical quantum theories of subatomic systems represent dynamic variables by abstract mathematical operators possessing properties that specify the behavior of the system.

WHAT IT REALLY MEANS:

The primary goal of physics is to give a complete description of how the world works. Ideally, physicists would like a set of interrelated mathematical equations which could explain and predict the actions of phenomena as diverse as the vast pirouettes of galaxies, the pull of the moon upon the tides, and the way subatomic particles bind together to form the nucleus of an atom. In other words, physics seeks to describe what the universe is made of and to mathematically specify how its constituents interact.

Three hundred years ago, Isaac Newton watched that famed apple fall to the ground, wondered whether that might have anything to do with the moon, and was inspired to invent classical mechanics. Newton not only developed a "world view," he also developed a series of equations that worked exceedingly well in explaining the behavior of moving bodies on earth as well as the motions of heavenly bodies. Classical mechanics contains a very precise, verifiable set of predictions that serve to describe how the world works on what might be called "the level of common experience."

The tremendous scope and predictive power of Newton's theories yielded a picture of the universe as a huge, complex, but theoretically predictable machine. It was believed that if only we knew in complete detail the initial state of the cosmic machine—the positions, interactions, and energy states of the component parts—we would then be able to predict all future events. This clocklike image of the universe governed science and industry for centuries and did not begin to collapse until experimental scientists developed tools to probe the world of the very small. Much to the dismay of physicists, Newton's equations seemed to lose their predictive power when they were applied at the atomic level. The theory that developed to replace them not only involved a new set of equations, it also represented a totally new way of looking at the world.

Quantum theories hold that four features of our common experience are illusory on the most fundamental level. First, classical mechanics and common sense tell us that a wave is a wave and a particle is a particle and that we can devise ways to tell the difference. Quantum theory teaches that there is *no* distinction between a wave and a particle except the method we choose to measure it. Particles can behave like waves and waves can act like particles. It depends entirely on how they are observed.

The second common-sense assumption of Newtonian-era physics is that there is an objective reality apart from our minds and that this reality can be observed without changing it. Quantum theory holds that there is no reality apart from experience and that the act of experiencing alters reality.

Thirdly, common experience leads us to believe that when we measure an object, no matter how small it is, we can determine both where it is and how fast it is moving. Because of the uncertainty principle, quantum theory tells us that we cannot measure one of the properties without distorting the other. There is a fuzziness in the finest picture we can resolve of the real world, and the fuzziness is not in our technology—it is in the way we perceive.

The fourth detour from common sense and classical mechanics takes off from the old assumption that science can predict the outcome of specific events with total accuracy (e.g., the behavior of bodies in motion). Quantum theory says that the world is fundamentally indeterminate. It is not possible, on the subatomic level, to predict exactly what will happen in any circumstance. The best one can do is to give an estimate of the *probability* that an event will occur. The behavior of a large number of molecules, as in a volume of gas, can be accurately estimated, but the actions of individual molecules are unpredictable. Newton's clocklike cosmos is replaced by quasi-illusory probability waves.

These world-shaking concepts started with a couple of relatively simple experiments. According to data gathered using diffraction techniques, light behaves as only a wave phenomenon can behave—by exhibiting interference patterns. According to equally legitimate data obtained by studying the photoelectric effect, light also behaves as only a particle can behave—by acting like a stream of individual "bullets." This paradox about the nature of light constituted the opening wedge that destroyed classical physics and loosed the strange beast of quantum theory upon the world.

Out of the wreckage of Newtonian mechanics, two schools of thought emerged. Those physicists who thought the universe was composed of particles were ably represented by Albert Einstein. He won the Nobel prize for demonstrating that energy could be emitted in the form of light

Quantum Theory—Two Bright Ideas

particles, or *photons.* These discrete packets of energy were the *quanta* posited by Max Planck at the turn of the twentieth century. According to this viewpoint, the wavelike aspects of light were merely reflections of the changing probability of finding a particle at any given point along its stream of radiation.

The opposite school, represented by Erwin Schrödinger and his wave equations, demonstrated with equal authority that the particles scientists thought they detected were hypothetical nodes along standing waves of energy. He further proved that the wave and particle perspectives were mathematically equivalent. Niels Bohr took that equivalence to its ultimate conclusion with his principle of complementarity, demonstrating that wave and particle models are complementary aspects of the same reality. It all seemed unbelievable to physicists of that era: physics, which in the nineteenth century thought itself to be on the verge of describing the universe in a neat set of equations, was no longer sure about how to describe the fundamental constituents of reality!

When Werner Heisenberg demonstrated that the wave-particle paradox can only be resolved by taking the observer into account, physics was changed for all time. Science, in the form of physical experiments with light, had uncovered a metaphysical paradox. The search for an ultimate objective reality had dead-ended. While Newton's equations still worked on the level of common sense, our means of learning about the world was irrevocably altered by this discovery of its limitations.

CONVERSATIONAL TACTICS:

If you're bold enough to tackle quantum physics in casual conversation, be prepared for cynical feedback. Even though fifty years have passed since the core of the theory was first advanced, it seems just as radical today. When the most common tech-talk put-down—"You don't know what you're talking about"—is aimed your way, tell the skeptic that there is no "what" to talk about and back up your claim by recounting the famous Copenhagen Convention.

In 1927, quantum theory was driving its creators so wild that they decided to meet in Copenhagen to decide what it all meant. They talked of constants and functions and drew Greek letters on blackboards, of course, but an important part of the theory stemmed from the metaphors the physicists used in their attempts to describe this new world, which could not be pictured in any of the old ways. The meaning of the mathematics seemed to do weird things to the minds of the people who developed the theory. They started describing reality in peculiar terms.

It was a pretty big issue at stake in Copenhagen, even though it was couched in esoteric jargon. If the equations were interpreted in one way, science could never hope to penetrate to an absolute reality. We would never be sure if the entities we were discussing were real waves and particles or artifacts of the way mathematics affects our minds. The Copenhagen interpretation proclaimed that findings on the subatomic level indicate that physics is *not* about any "real world" but about systematic correlations between our observations. The best we can do regarding the actual behavior of any particular particle is to specify a range of probabilities. The probabilities only become actual when we observe, and thereby alter, the particle. In one sense, this means that physics is ultimately the science of the structure of consciousness as much as it is the gauge of the universe.

The Copenhagen interpretation, quasi-mystical as it may be, is not the only way to skin an equation. The Everett-Wheeler-Graham hypothesis, also known as the "Many Worlds Interpretation," takes the issue into science-fiction territory. In quantum mechanics, an elementary particle is represented by a mathematical term known as a wave function, which yields a description of the moving particle as a blur of probabilities smeared over a region of space-time. When an observer conducts an experiment to find out whether the entity is here or there, a wave or a particle, the wave function of probabilities collapses to zero for all but one of the possible configurations. In plain English, if you discover that your experiment yields a wave, the chance that it is a particle becomes zero.

The Many Worlds Interpretation says that the moment of observation splits the entity into two different entities in two different universes! If the observer finds a wave at a specified time and place, then another universe must exist in which that entity is a particle. Worlds branch off worlds with every experience, each cosmos as "real" as every other one. Everything that is possible happens. Only our attempt to make sense out of "our world" creates the illusion of continuity. Physics is our detailed map of that illusion.

RELATED TOPICS:

Blackbody Radiation, Neutrinos, Particle Accelerator, Quarks, Relativity, Uncertainty Principle.

FURTHER READING:

The Dancing Wu Li Masters, Gary Zukav, William Morrow, 1979.

The Forces of Nature, P.C.W. Davies, Cambridge University Press, 1979.

QUARKS

DEFINITION:

The generic name for the family of subnuclear particles now thought to be the primary constituents of all other matter.

WHAT IT REALLY MEANS:

One of the major goals of science is to discover the ultimate building blocks of the universe; since 1963, *quarks* have been the best candidate. You would think, judging from their past record, that physicists would begin to get the idea that there may be no end to their quest. First, in the early nineteenth century, the old Greek idea of indivisible *atoms* was revived. A hundred years later, it was theorized and experimentally confirmed that the "ultimate" components of matter were actually built of yet smaller building blocks which came to be known as *protons, neutrons,* and *electrons.* More recently, it has been discovered that even these fundamental particles can be shattered into over two hundred types of subparticles named *hadrons.* Now scientists are postulating that all of these sub-subparticles are made of even tinier things called quarks.

Quarks, for the most part, are what philosophers of science call theoretical entities. This is because these elusive subnucleons have great explanatory power in the context of physical theory, but there is not as yet an overwhelming amount of direct evidence that they actually exist in nature. One of the reasons such mythical beasts are taken seriously is that antimatter and neutrinos were also once theoretical entities; like quarks, these other strange forms of matter were originally introduced in order to tidy up some equations and were only later detected by physical experiment. Quarks are still very useful at tidying up equations but are proving much harder to find in nature.

Murray Gell-Mann and George Zweig first proposed a family of three quarks named *up, down,* and *strange* to account for the quantum characteristics of all nuclear particles observed up to that time, to correctly predict the existence of a nuclear particle not yet detected, and to prohibit the existence of other nucleonic states not found in nature. Quark theory has been greatly elaborated since 1963, and the family of quarks has grown to six with the addition of *charmed*, *top*, and *bottom* quarks.

Whether and how quarks might move from a theoretical to an actual status is much debated. While the existence of quarks is becoming more and more acceptable for physicists as an indirect consequence of recent experiments regarding their believed properties, many scientists now feel that quarks are so tightly bound to one another that it may never be possible to break one loose for direct examination. If a quark is ever forced out of a hadron, physicists may find once again that so many new subquarks are released that their status as primary building blocks will be called into question. When that happens, physicists will finally have to decide whether there are any basic constituents of all matter, or whether the quest for the infinitesimal can continue indefinitely. Either answer promises to be puzzling in ways we can't yet conceive.

CONVERSATIONAL TACTICS:

Not only are there "strange" quarks, but quarks as phenomena are all pretty strange. It is even stranger to converse about them. What can you really say about a group of objects referred to as "flavors," as in "That quark is the fourth flavor, charm"? Physicists now also believe that quarks come in three different "colors"—red, green, and blue—although this designation bears no resemblance to the colors our eyes can see. If all this seems a bit too artlike for tech talk—almost literary, in fact—you're on the right track.

The term *quark* was taken from a line in James Joyce's *Finnegans Wake:* ". . . three quarks for Muster Mark." It seems fair to say that the average scientist has no more understanding of quark theory than the average reader has of Joyce's enigmatic novel.

If the literary approach isn't sufficient, shift the discussion from the theoretical to the technological by mentioning that high-speed printing technology is the newest hope for a physical means of detecting quarks. Two scientists at Lawrence Berkeley Laboratory, Greg Hirsch and Ray Hagstrom, read an article about ink-jet printing in *Scientific American* and realized that this technique could be adapted to the quark search. In ink-jet printing, a fine stream of ink is squirted from a nozzle and broken up into tiny drops, each one of which is given an electric charge. The drops are steered by a changing electrical field to form the shapes of letters.

The new quark-hunting method involves a similar stream of small drops of mercury. As the drops pass through the electrical field, each one is deflected by an amount proportional to its electrical charge. Because electrical charge is quantized rather than continuous, the stream of drops spreads out like a fan, with separate "fingers" representing a particular integral of a charge. If any drop contains a single quark, then a fractional charge will be imparted, so that drop should land between the fingers. So far, it sounds like a good idea, but the final results are not yet in.

RELATED TOPICS:

Neutrino, Particle Accelerator, Quantum Theory.

FURTHER READING:

The Particle Play: An Account of the Ultimate Constituents of Matter, J. C. Polkinghorne, W. H. Freeman, 1979.

"A Gaggle of Quarks," *Frontiers of Science—II,* Government Printing Office, 1979.

QUASAR

DEFINITION:

Acronym for the term *QUASi-stellAR radio source.* These are very remote, small, luminous celestial bodies which emit strong radio signals.

WHAT IT REALLY MEANS:

One major task of astronomy is extending the known boundaries of the universe. A major sensation was caused sixty years ago when it turned out that some of those faint, starlike dots of light on the astronomers' photographic plates were actually other galaxies—unthinkably distant clusters of stars grouped in gigantic archipelagoes even larger than our own Milky Way. Today, astronomy no longer relies solely on visual sightings. Radioastronomy was developed in the 1950s, and it didn't take long before astronomers used it to further enlarge the limits of space by identifying a new and shockingly distant denizen of the sky—the quasar.

Radio telescopes "see" a different image of the sky than does the optical telescope. By matching radio sources with visible objects, astronomers have been able to give more complete descriptions of many celestial mysteries such as pulsars and supernovae. The quasar, however, retains its mystery. They were first discovered in 1960, when radio astronomers found intense sources of radio signals that had very small diameters. Further work demonstrated that while quasars emit trillions of times more radio energy than any known star, they are millions of times smaller than even the smallest galaxy.

Add one more enigmatic piece of information to the quasar puzzle: Spectral analysis of quasar light indicates that it is caused by hydrogen emission, the same as stars and galaxies, but the light patterns are shifted extraordinarily far down to the red end of the spectrum. This means that quasars must be moving away from the earth at incredible speeds, up to 90 percent of the speed of light, and some of them appear to be as far as fifteen billion light years from earth. The quasar light we now view through telescopes began its journey very soon after the creation of the universe. How could such a small, distant object shine such a powerful beacon through all that space and time?

Astronomers trying to explain quasars are on the horns of a classic scientific dilemma. If quasars were closer than scientists now think they are, there would be no need to postulate such incredible incandescence. However, this would mean that the phenomenon of the red shift, a cornerstone of modern astronomy, has to be incorrect. If the red shift data are correct, then another explanation for quasars' potency is necessary. What physical force could cause such a release of power? One group of astronomers proposes that quasars are yet unexplained events which took place long ago in the center of young galaxies, perhaps great bursts of energy released when the galaxies condensed. Other proposed explanations include black holes, white holes, matter-antimatter galaxies in collision, and chains of supernovae.

Whatever their nature and origin, quasars are excellent candidates for the "city limits" markers of the universe. No quasars have been found beyond the fifteen-billion-light-year limit, and this may be as far back into time as we will ever be able to look, because the lights of the heavens were not "switched on" until that time.

CONVERSATIONAL TACTICS:

Since the true nature of quasars is presently unknown, conversing about them can get pretty tricky. There is, however, an alternative conversational strategy that can be a real showstopper if you do it correctly: Talk about "quasar," the word, rather than the astronomical object. *Quasar* is an acronym, a word formed from the initial letters or groups of letters of words in a phrase—in this case, from the phrase "quasi-stellar radio source." Science is full of acronyms, but almost everyone has forgotten the original phrase. This list of origins will enable you to decode the following acronyms for your audience:

ASTRONOMY:

Parsec *Par*allax *sec*ond, an astronomical unit of length, equal to 3.3 light years.

PHYSICS:

Laser *L*ight *a*mplification by *s*timulated *e*mission of *r*adiation, a device for producing an intense beam of coherent light.

Radar *Ra*dio *d*etecting *a*nd *r*anging, a method of characterizing an object by the reflection of radio waves.

COMPUTERS:

Bit *Bi*nary digi*t*, a unit of information-storage capacity.

FORTRAN *For*mula *Tran*slation, an early programming language.

CAD/CAM *C*omputer *a*ssisted *d*esign and *m*anufacture, a process for increasing automation using computers.

MISCELLANEOUS

Rand *R*esearch *an*d *d*evelopment, the well-known think tank.

Scuba *S*elf-*c*ontained *u*nderwater *b*reathing *a*pparatus, or diving equipment. Also known as the aqualung.

MAD Mutual *a*ssured *d*estruction, unbelievable as it sounds, is the name for our current nuclear war strategy.

RELATED TOPICS:

Big Bang, Cosmology, Galaxy, Radio Astronomy, Red Shift, Stellar Death.

FURTHER READING:

Cosmos, Carl Sagan, Random House, 1980.
Galaxies and Quasars, William J. Kaufmann III, W. H. Freeman, 1979.

RADIO ASTRONOMY

DEFINITION:

The science of observing celestial objects by analysis of emitted or reflected radio-frequency waves.

WHAT IT REALLY MEANS:

The night sky offers an awesome and useful map of the universe to the most highly evolved optical sensor on earth—the naked eye. For millennia, human eyes and minds have attempted to navigate the seas, map the heavens, and time agricultural cycles by watching the motion of those lights in the sky. In the sixteenth century, the refracting telescope magnified human perception of the skies and forever changed our image of our place in the cosmos. More recently, astronomers have begun to use the discoveries of physicists and communications engineers to expand our window on the universe. Once again, our picture of the scheme of things—and our role in it—is changing dramatically.

We have learned since the invention of the radio telescope that our universe is much richer in information than we had ever dared to believe. We now know that visible light is but a thin slice of a much broader spectrum. We peek at the world through a tiny keyhole: the heavens are dressed in blue and white and red, but they are also cloaked in ultraviolet and infrared, in microwaves, radio frequencies, X rays, and gamma rays. Our unaided senses are tuned to but a fraction of the messages radiating through space. The history of Western science has seen a progressive expansion of our narrow sensory aperture. Our picture of the universe has been expanded by our instruments, from the electron microscope to the radio telescope.

We tend to think of radio waves as something new, but the cosmos has been in the radio business since the beginning of time. Heinrich Hertz showed us how to propagate radio waves in a laboratory in 1881, Guglielmo Marconi developed a working transmitter around the turn of the century, and we have been chattering away to ourselves (and everyone else in the universe) ever since. But it was not until the 1930s that anyone turned a sufficiently sensitive receiver to the heavens to hear what the rest of creation might have to say.

The man who first heard the biggest broadcast of all time was looking for something more practical than radio waves from outer space. His name was Karl Jansky, a young Bell Laboratories engineer in 1931, the year he happened upon what may be the most significant discovery in the history of stargazing. His objective was to find the causes of interference in transoceanic radiotelephone communications, so he set up an array of unshielded cables in an open field. In order to pinpoint sources of static, he moved the antennae by means of airplane wheels attached to a wooden frame.

Jansky quickly determined that most of the noise came from lightning, but he could not explain one persistent source of radio hiss that was independent of weather conditions. When he traced the location of the competing signal, he discovered that it was coming from somewhere among the stars. He consulted a star chart and found that his source of static coincided with the center of our galaxy. Although Jansky's field of interest was in communications, his findings were not lost on a few future-thinking astronomers, who realized that radio waves could pass through clouds of interstellar dust and gas which hide a great deal of the universe from our sight.

The big breakthroughs in the modern science of radio astronomy came after World War II when British radar research provided vast new fields of knowledge, as well as suitable surplus equipment. The theoretical framework for radio astronomy's experimental discoveries had been developed independently by a Dutch and a Soviet astrophysicist in 1944. Based strictly on calculations concerning the known properties of matter, they announced that hydrogen atoms in space—the basic stuff of stars and interstellar gas clouds—should emit radiation at a wavelength of 21 cm. This hypothesis served as an explanation of the mechanism by which entire galaxies could broadcast radio signals; equally importantly, the 21 cm prediction served as a calibration benchmark for observational studies.

In 1951, Harold Purcell and Harold Ewen at Harvard built a wooden antenna horn the size of a bathtub, lined it with copper foil, and stuck it out a window of Harvard's physics laboratory. They succeeded in receiving the 21 cm radiation and confirmed that it was most strongly detected when the antenna was aimed toward the heart of our galaxy. Radio astronomy grew to maturity in the 1960s and 1970s, providing a cornucopia of data for the theoreticians who were detecting such strange and tantalizing extraterrestrial entities as quasars, pulsars, and radio galaxies. Theories of cosmology and stellar death have benefited from data gathered by radio telescopes.

CONVERSATIONAL TACTICS:

Now that you know the basics of radio astronomy, *Talking Tech* will provide you with three bits of conversational ammunition—a look at the latest news, an amazing fact to open or close discussion, and something to think about. Because of two technological advances, the latest news from outer space should start coming thick and fast during the next five years. The Very Large Array (VLA) radio telescope in Socorro, New Mexico, will furnish the most sensitive radio "ears" ever built, and orbiting observatories are opening up two new windows on the electromagnetic spectrum by creating the fields of X-ray and gamma-ray astronomy.

The amazing fact is that all the wide-ranging theories and impossibly distant phenomena opened up by radio astronomy are based on signals so faint that they collectively total less energy than that expended by a single snowflake when it hits the ground!

The something to think about has to do with the fact that radio astronomy might never have progressed further than Jansky's rickety arrangement if it were not for the urgency of war. Radar and radio astronomy were not the only technological offspring of World War II. The science of cybernetics and the computer revolution, the jet plane and the moon rocket, nuclear energy and microelectronics, synthetic fuels and antibiotics, were all either created directly or spurred indirectly by war research. In this sense, much of modern science and technology was literally built out of war surplus.

RELATED TOPICS:

Big Bang, Black Holes, Cosmology, Galaxy, Image Enhancement, Quasar, Stellar Death, Sunspots.

FURTHER READING:

Frontiers of Astronomy, Scientific American/W. H. Freeman, 1970.
"An Ear on the Universe," John Lees, *Creative Computing*, March-April, 1976.

RANDOMNESS

DEFINITION:

A concept applied to groups or the processes which generate them, possessing the characteristic that any member of the group has an equally probable chance of being selected from the population.

WHAT IT REALLY MEANS:

Randomness, like many of our most fundamental concepts, is easier to understand intuitively than it is to explain in words. For example, in the above definition, *random* is defined in terms of "equally probable chance." But how is one to define this latter concept without appealing to the concept of randomness? The problem is that mathematicians have tried to give an absolute meaning to a concept which can only be defined relativistically. Such attempts always end in circularity.

The key to properly defining *randomness* is to yield to our intuitive understanding: Randomness is the absence of order, and order is person-specific. What may appear to be a random sequence of nine numbers to you, may be recognized by someone else as his Social Security number. In fact, some of science's greatest achievements have been nothing more than the imposition of order on a previously chaotic array. Thousands of years ago, 3141592653 was just a meaningless jumble. Today, it is recognized as the first digits of pi, the ratio of the circumference of a circle to its diameter.

The whole point is that there is no such thing as the absence of order; the absence of one pattern demands the presence of another. The most we can say about a so-called random sequence is that it displays a pattern which has not yet been interpreted. Randomness and order are determined solely by the perspective of the observer.

Not that scientists are forbidden to use the concept of randomness, but the concept must always be understood to be relative: A choice, sample, or event is random with respect to variable X if it is not significantly affected by variables that significantly affect X. In other words, we can use the roll of a die to determine the outcome of a game, since the die is not under the influence of any of the players; with respect to the players, the roll of the die is a random event. Scientific randomness is really pseudorandomness—random enough for the purposes at hand.

This definitional limitation has done nothing to limit the utility of the concept. As demonstrated below, randomness is a useful concept to many different brands of sciences:

Random access: Suppose you have a tape with all your favorite songs, and you want to hear a song that happens to be in the middle. How can you listen to that song without hearing all the rest? An analogous problem puzzled the early computer designers. It was easy to build a computer with an enormous memory by storing information on magnetic tape, but it was difficult to have the computer find the needed information from that memory quickly enough; it took too long to run through the tape until the needed portion was located. The solution was the construction of what are now known as random access memories (a.k.a. RAM): Instead of storing information in a linear sequence, information would be stored in a magnetic or electronic matrix. In this way, each bit of information has a specific address (column 47, row 93, file 6) which can be located and retrieved without "visiting" any other "address." Random access refers to the computer's ability to locate directly a specified piece of information stored in memory.

Random sample: Science is often characterized as the act of making general statements based on the evidence of specific instances; this is what is known as inductive reasoning. How can the Nielsen ratings "know" what one hundred million households are watching on television based on the evidence of only one thousand households? They have cleverly constructed a sample which reflects the population as a whole. In the Nielsen case, the sample is not random; rather, each member is selected in accordance with some preset criteria. There are, however, many instances in which using a random sample—a sample in which the members are statistically independent of each other—is the most efficient procedure: testing products off an assembly line, selection of draft inductees, and measuring the effects of an experiment on a large population. Often, however, randomness is a difficult commodity to obtain—the first three random number tables were subsequently discovered to be nonrandom!

Random walk: Imagine that as you leave your front door you flip a coin each time you want to take a step. If the coin turns up heads, you take one step forward; if it's tails, you take one step backward. How far, on the average, do you think you're likely to get? Since the coin toss is a random event, you should stay right around your doorway. We do know, however, that because of fluctuations as the number of tosses increases you will eventually get farther away from home. This silly-sounding game (a random walk) turns out to be an accurate mathematical model of the way atoms move in a gas—called *Brownian motion.* According to some statisticians, it is also a model of stock market behavior; they believe you are just as likely to hit it rich on Wall Street by throwing darts at the stock page as by studying trends and analyses.

CONVERSATIONAL TACTICS:

According to Warren Weaver in his book *Lady Luck,* a quirk in the distribution of the first digits in any table of numbers provides one of the greatest sucker bets of all time. The quirk is this: In any table of random numbers, or one you suppose might be random—things like population figures or motor vehicle registrations—the odds that the first digit of any number in the table is a 1, 2, 3, or 4 is 0.699. With this knowledge, you can proceed to make the following bet: You allow a friend to open any book that might contain tables of supposed random numbers (*The World Almanac*) to any page and to pick twenty numbers according to any scheme desired (the first twenty digits, every other number, etc.).

Since you are generous, tell your friend you will get one dollar for every number that begins with a 1, 2, 3, or 4, and your friend will be paid off on 5, 6, 7, 8, or 9. In other words, you are giving your friend odds of 5 to 4. The bet is a sucker bet because you know the odds are 7 to 3 in your favor. This all seems incredible and counterintuitive, yet Weaver provides a simple proof.

RELATED TOPICS:

Entropy, Information, Mathematical Modeling.

FURTHER READING:

How Real Is Real? Paul Watzlawick, Vintage, 1976.
Lady Luck, Warren Weaver, Anchor, 1963.

RECOMBINANT DNA

DEFINITION:

A molecular biological technique in which genetic material from one organism may be inserted into the chromosomal structure of a different organism.

WHAT IT REALLY MEANS:

Biologically speaking, the 1980s are the most important years in the entire four-billion-year span of evolution. While it is true that a few human inventions of lasting importance were discovered in the past—agriculture, the wheel, the computer—even these world-changing technologies didn't possess the revolutionary power now in the hands of the molecular biologists. When Watson and Crick deciphered the alphabet of the DNA molecule thirty years ago, human evolution first became truly conscious of itself. The outcome of their discovery was more than pure knowledge—it also put us very close to unprecedented power. For the first time on this planet, a living creature has the capability to manipulate the global evolutionary process. The technologies which are already flowing from basic discoveries in molecular biology are creating a world radically different from everything that has gone before.

Recombinant DNA—The New Mythical Beasts

This infant science already promises to irrevocably alter our *material* well-being—for good, and possibly for ill. It also poses serious questions about our *spiritual* and ethical well-being. This is the paramount example of a scientific matter that could become too important to be left to the scientists. Poor old Dr. Frankenstein's bioethical dilemma was a minor problem compared to the questions stirred up by present-day biotechnology. Just as Galileo's telescope was the technical trigger of the Copernican revolution, a specific laboratory technique—recombinant DNA experimentation—is the fuse of the current explosive controversy.

Like many other revolutionary ideas, the mechanism of recombinant DNA was tricky to work out at first, and is relatively easy to understand now that the details have been elucidated. Watson and Crick explained in their famous 1953 paper that the double-helix-shaped DNA molecule is the physical instrument virtually every organism uses to pass on genetic information; the structural characteristics of these magic molecules make them capable of reproducing exact duplicates of themselves as each cell divides. Furthermore, each DNA molecule is a "blueprint" for specifying the structure and order of manufacture of the proteins necessary for each life form. Because of the special abilities of DNA, every cell in your body contains the same genetic material as every other cell, and it is precisely this material which determines your form and shape. Find a way to alter that genetic material and you've found a way to make an organism to your own specification—to create life to order, even if that life is evolutionarily nonviable.

Only twenty years after the Watson and Crick paper, scientists had learned how to create life forms which had never before existed. Technically, it's called "the construction of biologically functional plasmids in vitro," but it really is the creation of new forms of life. It's done through the technique of recombinant DNA, also known as "gene splicing." Working with organisms much simpler than humans, principally bacteria called *E. coli,* scientists begin by isolating the bacterium's DNA; this DNA, which is called a *plasmid* and is ring-shaped, is then treated with another special compound called a *restriction enzyme,* which chemically "cuts" the DNA at a specific site, allowing the closed ring to open.

At the same time the plasmid-snipping takes place, a different organism, referred to as the donor, has its DNA blueprint altered in a similar way. The donor's slice of DNA is inserted into the opened plasmid, and an annealing enzyme is added to close the new hybrid ring. The doctored plasmid (known as a *chimera,* after the mythical hybrids of ancient Greece) is inserted back into the host bacterium, and the bacterium now possesses recombined DNA that replicates along with each successive generation of bacterium, at the exponential rate for which bacteria are so famous. If you leave one viable bacterium in a petri dish and return a few days later, you'll find billions of genetically identical bacteria.

The fragment of donor DNA may contain a gene for antibiotic resistance, or instructions for producing a specific protein. If your happily proliferating bacteria happen to have inserted instructions for manufacturing insulin, for example, then all you have to do to produce insulin is to keep your bacteria well-fed and rapidly reproducing. It is the combination of these two factors—the power to insert foreign instructions into bacterial chromosomes and the ability of the bacteria to then rapidly reproduce those instructions and carry them out—that lends this technique such great potential usefulness.

According to Aristotle, there is nothing in your medicine cabinet that can't be used to either help cure you or kill you. The question now is whether the promised benefits of recombinant DNA technology outweigh the perceived risks. Those speaking for the benefits have strong pragmatic arguments. In the words of Nobel laureate David Baltimore: "Anything that is basically a protein will be makeable in unlimited quantities in the next fifteen years." In the medical field, this means an abundance of previously precious substances such as insulin, interferon (a chemical produced by the body to fight viruses and possibly to combat cancer), human growth hormone (which may be used to help cure growth disorders, heal broken bones, treat serious burns, and perhaps slow the aging process), antibiotics, and enzymes. The production of large quantities of human hormones is already beginning. Eventually, recombinant DNA techniques might be able to directly cure genetic diseases by incorporating reengineered DNA into an individual's defective gene structure.

The nonmedical potential is just as staggering. Some experts estimate that the technique may eventually reduce capital, energy, and related costs of producing chemicals and pharmaceuticals by as much as 50 percent. Other scientists are now working to create new types of microorganisms with valuable talents—synthetic creatures able to manufacture natural foods, scavenge pollutants, mine minerals, and concentrate scarce materials. Besides converting microorganisms into miniature producers of commercially desirable substances, recombinant DNA may also be used to introduce useful characteristics into higher plants and animals. Work is being done to transfer the genes of nitrogen-fixing plants to those which can't fix nitrogen naturally, thus creating new, self-fertilizing crops. Work is also being done to improve the food value and increase the productivity of crop plants. Recombinant DNA research has spun off an entire "bio-industry" that may one day overtake other science-spawned industrial giants like IBM. Already, some recombinant researchers have become multimillionaires. Genentech, one of these new companies, created quite a stir on the stock exchange when it went public in 1980.

Initially, all these potential benefits were overwhelmed on the risk side by a deadly set of scenarios. The community of molecular biologists themselves first worried about producing an "Andromeda strain" organism, a doomsday bug that might be lethal to humans, possess no natural enemies, and be capable of reproducing quickly enough to wipe out the human race or overturn the whole evolutionary process. It sounds crazy, but theoretically this is not a crackpot fear. *E. coli*, the workhorse of the recombinant laboratories, is native to the human intestine. What are the dangers of putting a cancer-causing virus into an organism which could then settle inside human organs if it were to escape? If an antibiotic-resistant pathogen ever escaped the laboratory, would medical science be able to stop it? The scientists had a famous meeting at Asilomar, California, in 1975, to debate these frightening new possibilities.

Complex research protocols were developed, and the National Institutes of Health issued a series of new regulations. The city councils of Berkeley and Cambridge passed resolutions regarding the conduct of such research in their communities. In the late 1970s the legal status of recombinant DNA research was a political hot potato. Now, however, with a solid record of safety behind the procedures, the debate has quieted, the scientists have returned to their laboratories, and in January 1980, NIH relaxed its guidelines in order to facilitate further research.

If the argument against recombinant DNA had been simply a risk/benefit argument—material benefits versus risks to public health—it would be easier to claim that the benefits were more immediate and more important. A second argument was advanced, however, one that is less quantifiable but no less compelling to its adherents. This argument has a long history of being used against "scientific progress," to little avail, but in this particular case the objection may have considerable significance. It is simply the fear that there is a "forbidden knowledge" which humans are not wise enough to possess. There is no doubt that we now have it in our power to create new life forms. By the end of the century we will be able to genetically engineer human beings. Considering our track record—thermonuclear and biological weaponry, toxic chemical spills, the abuse of pesticides, the disappearance of entire species, and the erosion of crucial aspects of the environment—are we absolutely sure that we now have the wisdom to properly apply the powers of biotechnology?

CONVERSATIONAL TACTICS:

For billions of years, sexual reproduction was the only way to recombine genetic material. Then these smart-aleck scientists decided to do it all with glassware. Back in the early days of the debate over "high-risk" research, the theoretical similarities between artificial and *au naturel* genetic recombination almost resulted in one very strange research guideline. If your conversational companions are earnestly debating the pros and cons of restricting this kind of experimentation, tell them how close the NIH came to outlawing *sex.* Michael Rogers recounted the actual discussion in his book *Biohazard:*

> *Jane Setlow, from Brookhaven National Laboratories—then the only woman on the NIH committee—in the midst of examining the freshly copied Asilomar statement, started to laugh.*
>
> *Some of the other committee members stared at her across the polished table. "What is it?", Stettin wondered, after a moment.*
>
> *"Well," she said, "I just realized that, according to this, human sex has become a high-risk experiment."*
>
> *There was a brief silence, then laughter. Maybe, someone suggested, it should specify "genetic recombination in the laboratory."*
>
> *"But that still rules out sex," someone else said.*
>
> *Setlow glanced up. "Only," she said, "in the laboratory."*

RELATED TOPICS:

Altruism/Selfish Gene, Bioethics, Clones, DNA, Origin of Life, Reproductive Technology, Technology Assessment, Virus.

FURTHER READING:

Biohazard, Michael Rogers, Avon, 1979.

"Exploring the New Biology," *National Geographic,* Vol. 150, No. 3, September, 1976.

Who Should Play God? Ted Howard and Jeremy Rifkin, Dell, 1977.

RED SHIFT

DEFINITION:

The wavelength at which electromagnetic radiation from a distant object reaches earth is increased by the velocity of recession of the object with respect to the earthbound observer; therefore, light received from luminous celestial objects is displaced toward the lower-frequency (red) end of the spectrum.

WHAT IT REALLY MEANS:

Astronomy is a most peculiar science. First of all, it's notoriously difficult to conduct astronomical experiments. You simply can't control the time and place of a solar eclipse. Secondly, the data astronomers use is hundreds or thousands of years old, because it takes that long for the light of distant events to reach earth. Finally, even though astronomy is our most ancient science, only in the last few decades have astronomers forged the tools to explain as well as predict heavenly phenomena. These facts make the 1930 discoveries of a pair of American astronomers, Edwin Hubble and Milton Humason, even more fantastic: Employing a series of fuzzy lines on old-fashioned photographic plates, one big telescope, and a few equations, they were able to expand the boundaries of the known universe by ten billion light years.

Those fuzzy lines were actually *spectra*—the marvelous rainbows prisms produced from light. These spectra, however, were originally produced in distant galaxies. The observers were using a known principle of physics: When an element is burned in a laboratory, it absorbs specific light frequencies, so the light emitted from that flame has a series of shadows known as "absorption lines" on its spectrum; the pattern of lines is unique to the element burned. Since they also knew that the light from stars consists of burning hydrogen radiation, astronomers Hubble and Humason shunted the faint images captured by their telescope through prisms, and the resultant spectra were compared with laboratory-measured samples of burning hydrogen.

When the laboratory and extraterrestrial spectra were compared, a mind-boggling discovery emerged. It turned out that the hydrogen absorption lines from distant galaxies were not the same as the hydrogen spectra observed here on earth. The same characteristic pattern of lines show up on stellar spectra from outside our own galaxy, but they are shifted toward the red end of the spectrum. The explanation for this red shift is one of the foundations of modern astronomy.

Most elementary physics courses contain a page or two on the Doppler effect—the reason an approaching train's whistle sounds higher in pitch until it reaches the listener and then drops in pitch when it recedes. This phenomenon occurs because the time between sound-wave peaks increases as the object and observer move apart, making the apparent wavelength longer and decreasing the pitch. In a brilliant leap of analogical reasoning, the pioneer astronomers inferred that the red shift must be a visual counterpart of the Doppler effect: The light from a star which is rushing away from the earth has its wavelength "stretched out" like the sound waves from the train whistle, causing the absorption spectra to shift toward the shorter frequencies. By measuring the amount of red shift, it became possible to tell how far away and how fast the galaxies were receding. It turned out that the universe is vastly larger than anyone had previously dared imagine. More important, *all* galaxies seem to be red-shifted, which led directly to the theory that the cosmos is expanding as the result of a "big bang."

CONVERSATIONAL TACTICS:

If someone tries to get the tech-talk drop on you by expounding on the red shift, you should have a couple of tidbits ready to drop into the stream. "Ahh, the Balmer series," is always a good remark regarding the red shift—jargon for the unique absorption pattern of hydrogen. *Blue shift* is another good term to remember. In an expanding universe, which is implied by the observed red shift of all other galaxies, a few nearby stars will be drifting randomly *toward* earth instead of rushing away. Because the light-emitting object is traveling on the same path as its light wave train to earth, the frequency of the wave train (its whistle) decreases, pushing the absorption lines toward the blue end of the spectrum. It is always a good topic to discuss when your conversation is threatened by an approaching train, plane, or person on roller skates carrying a loud radio.

RELATED TOPICS:

Big Bang, Cosmology, Galaxy, Quasar, Radio Astronomy.

FURTHER READING:

The Red Limit, Timothy Ferris, Bantam, 1979.
Black Holes, Quasars, and the Universe, Harry L. Shipman, Houghton Mifflin, 1980.

REGENERATION

DEFINITION:

The ability of certain organisms to regrow severed body parts.

WHAT IT REALLY MEANS:

The artificial limb of the future might not be made of Teflon after all. A new school of medical research proposes that the ideal material for organ replacement is your own tissue. All those futurist scenarios about bioengineered limbs and synthetic heart implants may seem laughable to our grandchildren, since many scientists in this field guess that by the year 2000 you won't replace your worn-out heart, hip, liver, or lungs—you'll simply grow new ones.

There are distinct medical disadvantages to ripping open your chest cavity and inserting foreign substances. Wouldn't it be safer and more pleasant to activate the mechanism in your body that grew the missing part in the first place? Solid evidence from lower species supports the hypothesis that some combination of electromagnetic and hormonal stimulation can induce regeneration in *any* species.

Regeneration may be the bright biotechnology of the 1980s, but the roots of the field go back to an Italian physiologist, L. Spallanzani, who published his regeneration research in 1768! He furnished the first two important clues in the experimental puzzle: The younger the animal, the greater its regenerative capacity, and the lower an animal is on the evolutionary scale, the greater its capacity for regrowing tissue. Other research, done in the late eighteenth century, discovered that an electrical charge is generated at the site of an injury. This *current of the injury* turned out to have significant import when biomedical investigators took up the regeneration quest in earnest after World War II.

In 1945, American biologist Meryl Rose used a strong salt solution to induce regeneration in the amputated forelegs of frogs, thus proving that an animal which does not naturally regenerate can be artificially stimulated to do so. In 1958, Russian investigator A. V. Zhirmunskii found that the current of injury is proportionate to the amount of nerve tissue in the area. At that point, an American orthopedic surgeon, Dr. Robert Becker, fitted these few crucial pieces of evidence together into a startling new theory which, if confirmed, could become the basis for a science of induced regeneration.

In 1958, Becker began to think about the old idea of the current of injury. He knew that nerve tissue is related to regeneration, because lower orders of animal have comparatively more nerves in their extremities. He also knew that both the trauma of the injury and the presence of nerve tissue are related to the current of injury. It was then that Dr. Becker set out to discover whether the current of injury was related to regeneration.

By measuring the electrical characteristics of regenerating limbs in salamanders and comparing them to those of scarring limbs in nonregenerating animals, Becker discovered that the current faded away rapidly in nonregenerators but switched polarity and maintained strength throughout the growth process in salamanders. This pioneering work was followed up in 1964 by Steven Smith, who implanted simple batteries in the amputated forelimbs of frogs, putting the negative end at the stump, and induced regrowth in a species that did not regenerate naturally.

Research to date has confirmed the hypothesis that induced voltage can stimulate regeneration: Becker made a significant breakthrough in 1973 by electrically inducing forelimb regeneration in a mammal—a white rat—in just three days. Also in 1973, Becker's years of study into the electrical properties of bone paid off. Implanted electrodes speeded up the healing of fractures in human patients who had lost the ability to regenerate broken bone tissue. More recent experimentation has revealed a possible physiological explanation for the success of electrical

Regeneration—A New Reason for Slit Skirts

regeneration experiments. Although the nerve cells work on a digital, strictly on-or-off series of impulses, the special cells surrounding nerve cells appear to carry an analog current, which could act as a channel for electromagnetic regeneration.

CONVERSATIONAL TACTICS:

All the talk of electrodes and salamanders is hopeful news to those with imagination and can provoke speculation on the conversational plane, but the idea often meets resistance on the grounds that it's a long way, biologically, from a salamander to a human. You should hold the most recent data in reserve for this inevitable rebuttal. In 1974, British physician Cynthia Illingworth published her finding that many young children can regenerate almost the entire first joint of their fingers within three months of the injury. Becker's fundamental research on bone is also generating clinical research: Dr. Andrew Bassett and his colleagues at Columbia University are successfully using electromagnetism to promote healing in bone fractures.

While most of the debate is centered on regeneration of limbs, you can always refer to the work on regeneration of other, more vital organs. Russian scientist V. Polezhaev cut away the scar tissue on the hearts of dogs that had suffered severe heart attacks; all of the hearts regenerated tissue, and less than 5 percent of the dogs died.

RELATED TOPICS:

Acupuncture, Longevity—Immortality, Origin of Life.

FURTHER READING:

"The Miracle of Regeneration: Can Human Limbs Grow Back?" Susan Schiefelbein, *Saturday Review*, July 8, 1978.

"Bioelectricity and Regeneration," R. B. Bogens, *Bioscience*, August, 1979.

RELATIVITY THEORY

DEFINITION:

Two related theories, the special and the general, which deny the possibility of absolute motion and explain the behavior of all moving bodies by appealing to the concept of gravitational force.

WHAT IT REALLY MEANS:

The great British biologist J.B.S. Haldane said that the universe "is not only a queerer place than we imagine—it is a queerer place than we *can* imagine." Although he wasn't talking specifically about the special theory of relativity, there is no better example of a scientific idea capable of stretching imagination past the breaking point. The legend has it that young Albert Einstein mused about what it would be like to ride on a light beam, and this childhood daydreaming later led him to the theory of relativity. Try to imagine what it would be like to ride on a light beam: that simple fantasy took little Albert and all the rest of us a very long way.

Of course, good science develops out of more than a child's reverie. In the case of the special theory of relativity, it developed as a solution to two puzzles which had long been plaguing physicists. The first, a quasi-philosophical riddle dating back to the ancient Greeks, was easy to state: Is there any such thing as absolute motion? This question was of more than philosophical importance, because it was a linchpin of classical mechanics—the study of bodies and forces. The laws of classical mechanics, as developed by Galileo and Newton, worked exceedingly well, but they postulated that absolute motion (motion that is non-relativistic) was possible, in fact necessary, and that the reference point of this absolute motion was known as the *inertial frame*. Given an inertial frame in which all the laws of mechanics were valid, it was easy to prove that the laws were valid in all other frames of reference. Unfortunately, after three hundred years of searching, no one had yet discovered the hallowed inertial frame.

The second puzzle, of more recent vintage, was a consequence of the work of two American physicists, Albert Michelson and Edward Morley. In 1887, they conducted a series of experiments designed to determine whether or not the earth was enveloped in an *ether*—a propertyless substance thought necessary for the propagation of light waves. (If light is a wave, like an ocean wave, it must be waving *something*—an invisible ocean, an ether—the physicists decided, not unreasonably.) What Michelson and Morley found, instead of an ether, was that the speed of light was a physical constant. No matter how it was viewed, or its source accelerated, light always was found to travel at the same speed—186,000 miles per second.

The constant speed of light was a direct violation of the Galilean transformation laws, which state that velocities are always additive. You can easily prove this to yourself the next time you ride an escalator. Time the ride while you remain stationary on the escalator step; then time the ride while you walk up the escalator. To nobody's surprise, the second ride will be shorter, because you were moving faster. You have to add your walking speed to the escalator's speed to arrive at your actual velocity. Amazing as it seems, the Michelson-Morley result means that two beams of light going up that escalator get to the top *at the same time*, even though one light source is fixed and the other is being accelerated up the moving staircase!

Einstein's solution to these puzzles was simple and logical. He couldn't argue with experimental fact, so he *postulated* the principle of the constancy of light. For the sake of further mental exploration, he just assumed that this paradoxical principle was true and tried to work out the logical implications. Seeing that this principle was in conflict with the Galilean transformation laws and knowing that no inertial frame had ever been found, he dispatched the notion of absolute motion and decreed that all uniform motion was relative. These two easily understood tenets are the foundation o the special theory of relativity. However, what follows from them are some pretty unimaginable consequences.

From the perspective of an observer who witnesses the motion:

1. A moving object measures *shorter* in its direction of motion and *disappears* at the speed of light.

2. A moving clock runs *slower* as its speed increases and *stops* at the speed of light.

3. The mass of a moving object *increases* as the speed increases and becomes *infinite* at the speed of light.

4. Space and time are not distinct entities but are simply space-time—a continuum.

5. Energy and matter are not distinct entities but are transformations of one another, related by the equation $E = mc^2$.

The first two consequences are intimately related because there can be no measurement of length without involving time and no measurement of time without involving length. This will probably seem nonsensical at first: "I don't look at my watch when I use a ruler." However, after a little thought you should see that measuring with a ruler requires that you know where the front and back ends are located at the same instant of time. Relativity theory states that the ends of the ruler will appear to be at different places at the same time for two different observers, one whose frame of reference includes the ruler and another who observes the ruler moving uniformly.

For the observer with the ruler, it will continue to measure twelve inches. For the "stationary" observer, the ruler will appear shorter as its speed increases. At this point it is tempting to say something like: "I get it—relativity deals with 'length as measured,' which will appear different for different observers. The real length of the ruler always remains twelve inches." At that point you *don't* have it. The first sentence is absolutely correct. There is no "real length"; the concept of absolute length is meaningless. What Einstein did was break down the barrier between an objective world and a subjective observer. All we can know is what we observe, and this is always relative to our frame of reference.

Was it easy for you to imagine yourself riding on a light beam? If so, understanding time dilation should be a snap. Start by thinking of light as the transmitter of visual information, then suppose you started riding the beam just as the clock struck 10:00. The next beam, the 10:01, would immediately follow you but could never catch up to transmit its information. For you, riding on the light beam, time at your starting place would be frozen at 10:00. If you were only going at 90 percent the speed of light, the 10:01 beam with its new data would eventually catch up. However, to the "stationary" observer, it would appear to take more than one second, so from that frame of reference time would be slowed. Riding on the light beam, or moving at 90 percent the speed of light, everything would seem "normal" in your reference frame. Since time is a relative concept, terms like "sooner" and "later" can only take on local meaning—local to a specific frame of reference. What is sooner in one frame may be later in another.

Because the concept of mass is defined in terms of length and time, both relative concepts, mass must also be a relative concept. Mass is the measure of the amount of stuff an object contains. To measure it, we must determine how much force is needed to accelerate the object by a given amount—that is, how much force it takes to propel an object across a distance in a given unit of time. Since relativity demonstrates that, to a "stationary" observer, a uniformly moving object is shrinking in length and dilating in time, it will also appear that a greater force is needed to accelerate an object in that reference frame. Therefore, the object's mass will appear to increase.

Understanding the fourth consequence, that space and time are not distinct but form a continuum, requires a different feat of imagination. Since no one, Einstein included, can visualize four dimensions, the best one can hope for is conceptual comprehension. Here are three "thought clues" to get you thinking in the proper direction. First, you know from the discussion above that space (distance) and time are related by the concept of measurement. Second, you know from your own common sense that something cannot exist at some place without existing at some time, and neither can it exist at some time without existing at some place. Finally, there is the advice of the prominent American physicist J. A. Wheeler:

> *Time is really a length, not an independent concept. To appreciate the falseness of the usual distinction between space and time, consider the inconsistent use of feet to measure the width of a highway and miles to*

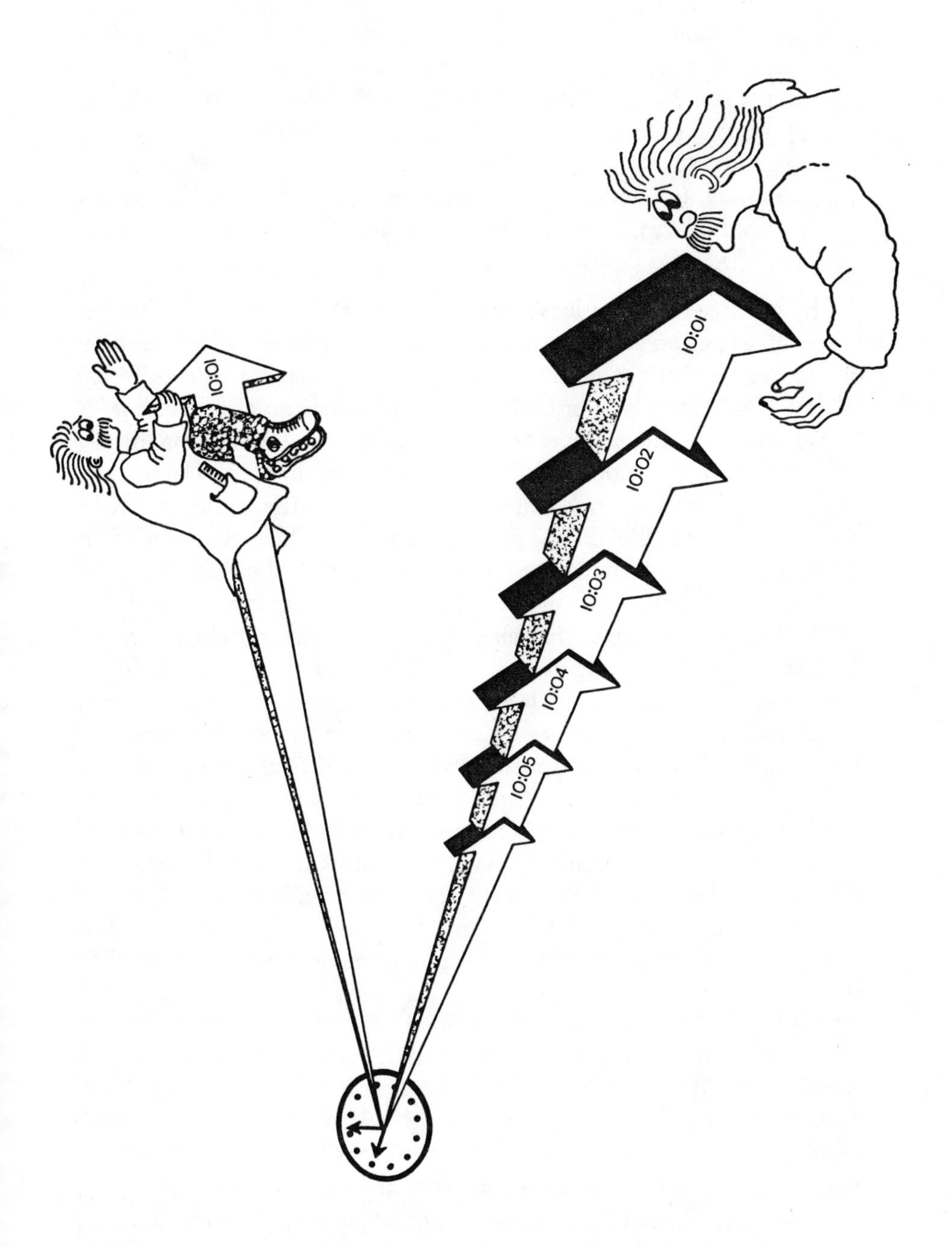

Relativity Theory—Young Albert's Dream

> *measure its length. Yet it is equally inconsistent to measure intervals in one direction in space-time in seconds, and to measure them in three other directions in centimeters. The conversion factor that changes one unit of length (cm) in the space directions to the other metric (sec), still of length, in the time direction is the speed of light—numerically three times ten to the tenth. And this factor is just as historical, indeed accidental, in character as the conversion factor 5280 that changes feet to miles! There is no more need to "explain" three times ten to the tenth than there is to "explain" 5280!*

The message of these clues should be clear. Prior to Einstein, human beings chose, as an act of conscious will, to perceive the world in three dimensions, with time as a separate concept. Although this conceptual scheme is still adequate for understanding phenomena at low speeds, it breaks down at higher speeds where relativistic phenomena occur. Furthermore, it is an outmoded and incorrect world view. Despite the difficulties in altering one's common-sense view of the world, it can be done (remember Columbus and Copernicus). The rigors of four dimensional space-time may be mired in mathematical abstractions, but an intuitive understanding of the concept is within everyone's grasp.

The fifth consequence, that mass is a form of energy and that energy has mass, is the most famous of all. It is also a direct result of the third consequence. When we apply force to a body, we start it in motion, which increases its amount of kinetic energy; but it also increases the body's mass. In other words, energy and mass are directly related; so long as the force continues, the energy and mass continue to increase. Because both mass and energy can be mathematically defined in terms of the concept of force, relatively simple calculus (no pun) demonstrates that mass and energy are equivalent according to $E = mc^2$.

The special theory of relativity started with talk about rulers and slow clocks and ended up with the H-bomb and nuclear power plants. The interconnectedness of scientific inquiry and events in the "real world" has seldom been so graphically illustrated.

The final refuge for those who find special relativity too mind-boggling and threatening is to proclaim that it is only a theory, not a fact. The issue of false distinctions aside, relativity is one of the best-confirmed theories of all time. Numerous experiments with elementary particles have unambiguously demonstrated that, from our earthbound point of view, the particles appear to live longer than the same particles at rest. This is because they are traveling so close to the speed of light that they undergo extreme time dilation.

If all the high technology needed for high-energy experiments evokes more skepticism, consider the 1972 experiment which only required eight clocks and an airplane. Four of the clocks, the most accurate ones available, were flown by plane around the world. When the plane returned, the clocks were found to be slightly behind their stationary earthbound counterparts, with which they were synchronized before the flight.

Because relativistic effects, like quantum effects, occur well outside our common experience, our individual survival is not jeopardized if we fail to embrace the new world view. It isn't like the law of gravity, which holds grave consequences for disbelievers. Yet our species survival is severely threatened each time we neglect scientific evidence and cling to old, outworn dogma instead.

If Einstein had retired from physics in favor of his beloved violin at this point in his career, his place in the history of science would have been secure—yet his efforts to restructure the way we all think about the universe were far from finished. In 1915, ten years after he published his special theory of relativity, the general theory appeared. According to most experts, it was an even greater intellectual achievement than the special theory.

Relativity Theory—Spaceship Earth

The special theory was "special" because it dealt only with uniform (nonaccelerated) motion. Einstein was not satisfied with such a limited case analysis. During those ten years, he was plagued by two persistent problems: Why should uniform motion be relative, while accelerated motion remains absolute? Isn't there one set of physical laws which is universally applicable to every individual's frame of reference?

He answered the first question by demonstrating that all motion is relative. In order to do this, he first had to prove his *principle of equivalence:* There is no way of distinguishing between uniform accelerated motion and a constant gravitational field. You can demonstrate the truth of this principle through a "thought experiment" of the type Einstein loved. Imagine that you're an astronaut about to be catapulted into space. At t-zero, you sense your body being slammed into your seat. The sensation may be explained in either of two ways: First, the earth remains stationary and the acceleration of the rocket causes the g force; second, the rocket remains stationary while the earth recedes, causing the same sensation of g-force. Since neither explanation has any prior claim to legitimacy, either may be correct; hence, all motion is relative.

The principle of equivalence also allowed Einstein to develop a set of physical laws which were invariant with respect to any observer. In other words, no matter how any observer is moving, the same set of physical laws may be applied. The key to these unifying laws was to replace any reference to bodies with references to gravitational fields. Einstein was able to do this mathematically by replacing the common sense idea of Euclidean 3-space with a non-Euclidean 4-space (the space-time continuum so popular with science-fiction writers). Although a complete understanding of these concepts requires knowledge of mathematical physics, an intuitive grasp may be obtained with some further visual imagining.

Picture a rubber sheet stretched out like a trampoline. Now place a bowling ball in the center of the sheet; it makes a large depression. Put a marble down near the ball on the rubber sheet; the small sphere rolls toward the larger, heavier one. According to Einstein, the ball isn't pulling the marble (gravity is not an attractive force, as Newton thought). Rather, the ball created a field which structured space in such a way that the marble, moving along the path of least resistance (also known as the *geodesic* of space-time), rolled toward it. In an analogous way, space-time is curved and warped by large masses like planets and stars, and this warping is what we call a gravitational field. The geometry of 4-space is simply an accurate description, through physical

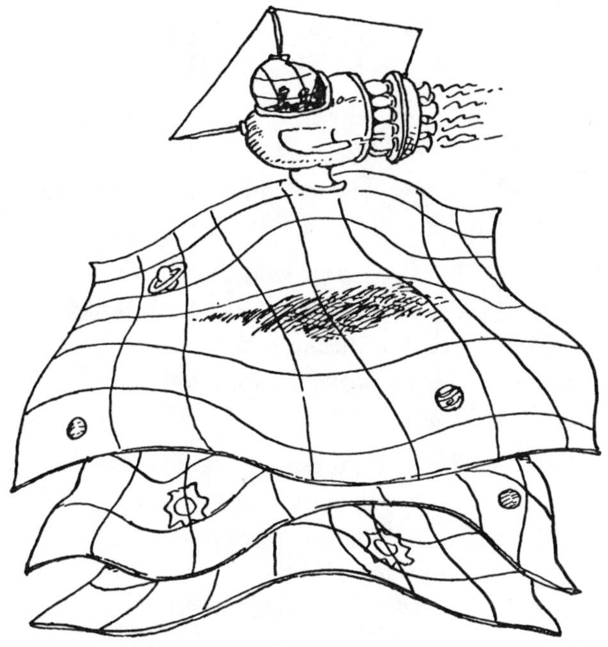

Relativity Theory—Warp Drive

laws and mathematical equations, of how bodies and forces interact.

Both the special and general theories of relativity are so incredible, their consequences so earthshaking, that they are difficult to believe at first reading. More likely, the discerning reader will pass through William James's three stages of acceptance: first, proclaiming the theories to be "false"; second, admitting that they are "true but trivial"; finally, finding them "not only true, but just what I believed all along."

CONVERSATIONAL TACTICS:

Most people think of science and poetry as diametrically opposed disciplines, yet the theories of relativity have an undeniably poetic elegance and simplicity about them. They have also inspired a good deal of poetry.

If you wish to pay homage to Einstein, you might recite these two epigrams:

Nature and Nature's laws lay hid in night:
God said, "Let Newton be!" and all was light.

—Alexander Pope

It did not last: The Devil howling "Ho!
Let Einstein be!" Restored the status quo.

—SIR JOHN COLLINGS SQUIRE

If you need a humorous way to explain the special theory, try one of these anonymous limericks:

There was a young fellow named Fisk
Whose fencing was exceedingly brisk;
So fast was his action,
The Fitzgerald contraction
Reduced his rapier to a disk.

There once was a lady called Bright
Who could travel faster than light;
She went out one day,
In a relative way,
And came back the previous night.

RELATED TOPICS:

Black Hole, Entropy, Fission, Fusion, Quantum Theory, Topology, Uncertainty Principle.

FURTHER READING:

The Dancing Wu Li Masters, Gary Zukav, William Morrow, 1979.
The Relativity Explosion, Martin Gardner, Vintage, 1976.

REPRODUCTIVE TECHNOLOGY

DEFINITION:

Any biomedical device or procedure used to alter, manipulate, hinder, or augment the human reproductive cycle.

WHAT IT REALLY MEANS:

Science has intruded on every other part of our lives, for better and for worse, so why did we think our most intimate human aspects were immune from the advance of technology? While we've been distracted with meltdowns and microcircuits, the glass artillery of experimental biology has been zeroing in on the last bastion of human mystery—sex and reproduction.

There are sperm banks in major cities around the world, and a little girl named Louise, now living in England, was literally conceived in a test tube. The abortion issue, the advent of global overpopulation, and recent advances in basic biological research are three of the many vectors due to intersect with our sex lives during the 1980s and 1990s. In the more distant future, twenty to fifty years from now, the worst nightmares of Huxley, the blackest dreams of Hitler, or hitherto undreamed-of bioengineered utopias may result from today's discoveries in this field. Along with nuclear weapons, pollution, and the greenhouse effect, contemporary inhabitants of planet earth may have to face the fact that the billion-year history of sexual reproduction can now be changed into something utterly unrecognizable.

The techniques of artificial insemination, in vitro fertilization, and the use of hormones or other chemicals to either prevent or promote conception are much more than the results of scientific research; they are also passionately debated ethical, legal, moral, and social issues as well as harbingers of those possible bioengineered futures. The degree to which we now understand the social impact of current knowledge will probably determine the fate of future generations. In five to ten years, when new discoveries and more highly perfected techniques face similar bioethical debate, the decisions we make on these issues today will be cited as precedent for even more crucial matters.

Sperm banks and *artificial insemination* are the most widely known, and by now the least controversial, of the new techniques. Maybe this is because they've been around the longest—the first human artificial insemination was done in England in 1790! Human sperm can retain viability outside the body of the donor if it is stored at very low temperatures. The thawed sperm can be introduced via syringe into the vagina of a fertile woman, resulting in an otherwise normal conception. The technology is simple, but the implications can be complex. The new field of bioethics is wrestling with questions concerning the legal status of sperm and ova and the rights of donors and recipients.

In vitro, which means "in glass," is currently one of the most fashionable Latin phrases in the reproductive technology business. In artificial insemination, only the sperm is artificially manipulated, and conception takes place within the mother's body. However, certain malfunctions of the fallopian tubes make some women incapable of conceiving, even though their ova are viable. Louise Brown, the celebrated "test-tube baby," was the first publicly acknowledged success at total in vitro fertilization. An ovum was microsurgically removed from Mrs. Brown's ovary, then transferred to a sterile glass dish, where it met Mr. Brown's spermatozoa. The resulting viable zygote (biotech jargon for a future human) was moved back into Mrs. Brown's body and brought to term.

Two truly chilling ideas lurk in the more distant future: *parthenogenesis* and *ectogenesis*. Parthenogenesis, also known as "virgin birth," would eliminate the need for males in the reproductive process because the egg does not have to be fertilized. Ectogenesis, the ultimate refinement of in vitro technique, would eliminate the need for both men and women, except as donors of spermatozoa and ova. Aldous Huxley thought his *Brave New World* was six hundred years in the future—but that was before the dawn of biotechnology. While neither of these techniques are likely to apply to humans for at least twenty years, current research does support the possibility of human parthenogenesis.

Rabbits, guinea pigs, and mice are known to develop spontaneous parthenogenic embryos—eggs which start to divide without fertilization from the male. A few of these embryos have been piggybacked on normal embryos and brought to term in laboratory experiments. There is evidence that ovarian tumors in humans may be derived from similar, short-lived parthenogenic embryos; true human parthenogenesis could conceivably happen as a spin-off from research into the causes of ovarian cancer.

Ectogenesis, in which spermatozoa and ova are brought together in a glass container and the embryo is brought to term in a totally artificial environment, is today seen as a much more remote possibility than parthenogenesis, at least for humans. However, the first attempts at artificial placentas and uteruses have already been tested in lower species. Should it become technically possible to breed one's own army in a warehouse full of glass wombs, who can doubt that an irresponsible head of state will plunge right in where bioethicists fear to tread?

CONVERSATIONAL TACTICS:

If you decide that direct frontal attack is your best conversational strategy in this explosive field, here is an opening remark guaranteed to be both scientifically accurate and socially startling: "Virgin birth is possible among turkeys." Turkey hens do, on occasion, lay viable eggs without the assistance of a tom turkey.

If you mention sperm banks and artificial insemination to your co-conversationalists, they're likely to think of them as elective procedures—something you might choose to do. Sperm banks are seen as a kind of insurance policy for men who desire a vasectomy. If recent scientific evidence is correct, sperm banks and artificial insemination may become the *only* ways males can continue to play a role in procreation! If this is a sobering thought, direct the discussion to the impending "sperm crisis."

A famous 1951 study found that the average sperm count of American men was about one hundred million per milliliter. In the mid-seventies, two different studies by reputable scientists appeared, each citing substantially lower figures—seventy-two million and twenty to forty million. In 1979, things really started to get serious. A Florida State professor not only did a sperm count study in which he found the lowest numbers yet—twenty million—he also found that toxic substances in the sperm were the cause of the low rate. Men of the current era may be facing nothing less than an epidemic of involuntary "chemical vasectomies."

RELATED TOPICS:

Bioethics, Clones, DNA, Endorphins, Pheromones, Recombinant DNA, Sociobiology, Toxic Chemicals.

FURTHER READING:

Who Should Play God? Ted Howard and Jeremy Rifkin, Dell, 1977.
A Matter of Life, Robert Edwards and Patrick Steptoe, William Morrow, 1980.

SOCIOBIOLOGY

DEFINITION:

A multi-disciplinary unifying scientific theory which claims that social behavior in all organisms, including human beings, has a genetic basis.

WHAT IT REALLY MEANS:

Science is the organized response to a universal human trait—curiosity. Our species seems to have an instinct for asking questions, for constructing experiments and then trying to explain the results. We seem to be curious about everything from the sand under our feet to the stars above us, but most of all, we are curious about *ourselves.* What is a human being? What is our purpose in the scheme of things, and just how did we arrive here, anyway? Why do we act the way we do? Are we the descendants of barely domesticated killer apes, or are we conscious beings with immortal souls? Why is sex so fascinating? What is love?

Great dramas and mediocre soap operas have been created in reaction to these questions. Religions have been born and wars have been fought over them. Now a mutant strain of biology is claiming that our most "human" traits, from poetry to adultery, from altruism to aggression, can be explained in terms of the molecular structure of our genes. To say the least, sociobiology is a very touchy subject. "Controversial" is a mild description.

"Survival of the fittest," although slightly inaccurate, is the phrase which sums up evolution to most people who had a high-school science course. These four words represent the essence of Charles Darwin's revolutionary intellectual achievement, the theory of natural selection. This revolution in biology was concerned with the acquisition of physical traits—how cats came to have claws, or how humans came to have handy thumbs. Sociobiology is now extending Darwinian concepts to account for the acquisition of *behavioral* characteristics. If the theory is correct, then all the animals on Noah's ark, including Noah, owed their sex drives, work habits, aggressive behavior, child-rearing practices, emotions, and morality at least partially to the survival tactics of their DNA codes.

In order to understand why everybody is so agitated about sociobiology, you need to remember a little (but not much) more biology beyond "survival of the fittest." Darwin based his theory on two indisputable facts: First, all organisms exhibit *variability,* a tendency for each individual to have slightly different traits from others of the same species, and for species to vary from each other; secondly, all organisms reproduce many more potential offspring than those which live to maturity. From these facts, Darwin drew two conclusions: (1) The environment selects those individuals best suited to survive and presents them with a greater chance to *reproduce;* (2) the characteristics favored by selection are passed on to future generations.

These two facts and two conclusions explain why giraffes have long necks (because those giraffes able to feed from the tops of trees are more likely to survive and produce more of their kind), and why blue-eyed people are not indigenous to tropical rain forests (the glare of the sun is too intense). One hundred and twenty years after the publication of Darwin's *On the Origin of Species,* it is scientifically well established that physical traits develop according to the laws of natural selection.

Sociobiology, as a distinct theory, is only about twenty years old. Thus far, the new theory has far less evidence to back it up than did the original physical evolution theory, but it promises to raise twice as much commotion as Darwin's. It was first proposed by a group of entomologists working with the social insects (bees, wasps, ants, termites), as a way of accounting for a specific behavior of individual insects—actions which seemed "suicidal." These theorists observed that certain members of a hive or colony engaged in self-destructive behavior performed solely for the benefit of others.

Insect altruism flew in the face of Darwin's theory. If you sacrifice yourself before you breed, your genes are losers in the evolutionary contest, so a gene for self-sacrifice should be selected out of the gene pool. Sociobiologists explained the survival of this behavior by proposing a theory of genetic selfishness: it could be genetically economical for an insect to sacrifice itself if it meant that its genes, through the similar genes of close relatives, had a better chance to survive and reproduce. This explanation led, in turn, to a more general theory that social behavior in all organisms might be explainable in terms of natural selection.

At first, no one paid much attention to the birth of this new subscience. However, in 1975, Harvard zoologist Edward O. Wilson published *Sociobiology: The New Synthesis.* As soon as the wider population found out about it, scientists and nonscientists alike were quickly drawn into furious debates over some of the theory's more controversial consequences. Sociobiology asserts that there are genetic differences between men and women which determine a large part of our sexual behavior, that in all kinds of behavior, including religion and love, human beings may owe as much to their genes as they do to their culture, and that the survival needs of our genes turn every social transaction into a kind of reproductive warfare.

Wilson's book made it clear that since natural selection works on the genetic level, behavior must be genetically determined to a certain measurable extent. He also extended the sociobiological analysis from insects all the way to *Homo sapiens.* Although he never said that behavior was *absolutely* determined by the genes—Wilson estimated that about 10 percent of our behavior is strongly genetically influenced—this heretofore obscure entomologist managed to stir up the nature/nurture, ape versus angel argument all over again. Battle lines formed over individual issues and over the general issue of whether an inquiry of such questionable ethical basis should even be allowed to proceed.

Take a look at the near and dear subject of human sexuality from the sociobiologists' point of view, and you'll understand why people get so exercised about this new idea cooked up by those insect scientists. In the first place, the two-sexes scheme exists because it introduces genetic variability—when an amoeba splits, the offspring have the same genes as the parent, but when sexual reproduction occurs, the new generation has a novel genetic mixture, inherited from both parents. In evolutionary terms, life is a game, not of aggression, but of *reproduction.*

The goal of every organism and every man or woman is to transmit the maximum possible number of his or her genes into the next generation. Look at it as a version of capitalism on the cellular level. The true basis of value is a trait specified by a section of the DNA code: the gene, the molecular unit of heredity, is the "money" of a sociobiological economy. Because they tend to survive and reproduce, and hence ensure the survival of their genes, those organisms which are successful at grasping the rules of the economy are winners. The only evolutionary wealth lies in having many descendants. From the point of view of the individual, the reproductive economy drastically affects one's approach to sexual behavior, depending on one's gender.

The human female produces, at most, about four hundred ova in a lifetime, only about thirty of which could possibly be brought to term. The human male, however, produces enough spermatozoa in one ejaculation to fertilize a billion women. To the male, parenthood is biologically finished at the moment of conception; the female only begins her biological involvement with conception. To the woman, eggs are expensive; to the man, sperm are cheap. The two sexes have evolved vastly different ways to spend their genes, according to the sociobiological view.

This genetic sexual economy, although consistent with the facts on lower levels of organism, has little scientific data to back it up in the human sphere. That present lack of experimental support doesn't stop the sociobiology theorists from using the conceptual scheme to explain all kinds of human behavior.

Where did the idea of a long engagement originate? A male's greatest disadvantage in the genetic survival game is the possibility of not knowing whose offspring he might be protecting—a cuckolded male is the biggest biological loser. However, by monopolizing a female's time, a courting male can wait long enough to make certain that the female is not already pregnant by a competing male.

Why the premium on virgins in so many cultures? Of course, they're not already pregnant, but there is another important reason—she has no previous sexual experience she might be tempted to repeat. Virginity is a sign of fidelity, and that is genetic money in the evolutionary bank to the winning male.

How did the sexual double standard come about? The sociobiological perspective is bound to generate controversy on this point alone.

Following the same reasoning as above, desertion is the female's equivalent to cuckoldry. The worst thing for a female's reproductive success is that her mate leave her to raise her young alone, while he goes out and drops his genes elsewhere. Knowing that her mate's sperm are cheap and his continued partnership valuable, females tend to tolerate a certain amount of "fooling around" as long as their mate continues to be a good provider. Females are genetically conservative, scrutinizing their potential mates' traits before allowing them to get too close. Feminists tend to dispute *this* idea.

Why is polygyny (one man with many wives) the predominant human mating arrangement? According to anthropologist George P. Murdock, of 849 human societies he cataloged, only 137 were monogamous, a paltry 4 permitted women to take more than one husband, and an overwhelming majority of 708 were polygynous. Since women (as the scarce resource) control the breeding market, how could this state of affairs come about? (Irven DeVore of Harvard claims that "males are a vast breeding experiment run by females.") The sociobiological answer is that polygynous cultures are reproductively advantageous to females because only wealthy and powerful males can afford more than one wife; hence, in these cultures, the females' offspring have a better chance to succeed. In Western cultures like ours, where polygyny is outlawed, we have developed a form of sequential polygyny in which wealthy older males often divorce middle-aged wives to marry younger women.

Sociobiology seems to have an explanation for everything from wasps' work habits to chimpanzee courtship rituals, from altruism in ants to homosexuality in *Homo sapiens*. Is it possible that one theory could unify such diverse phenomena? Only time and evidence will tell. Remember—it took Darwin's theory over a century to win general acceptance, and even today there is a large and vocal opposition. Sociobiology has only existed for twenty years, and already many of its explanations of insect behaviors are gaining acceptance. The real test remains to be settled—extending these doctrines to the higher animals and humans. The investigation, and the debate, may well continue into the next century.

CONVERSATIONAL TACTICS:

The most volatile charge against sociobiology is that it is politically biased and dangerously supportive of racist doctrines. If you want to launch an attack in this direction, you can start by quoting University of Chicago anthropologist Marshall Sahlins, who dismissed the field as a form of "genetic capitalism—an attempt to defend the current structures of Western society as natural and inevitable." A second attack, from the humanist perspective, is that we have evolved far beyond the determining power of our genes by becoming conscious and creating culture. Our rules, laws, language, and morality override the tendencies inherited from our more dangerous and aggressive past.

If you feel compelled to argue in favor of continued research in this direction, you might mention the claim of respected science writer John Pfeiffer that sociobiology is "an evolutionary event, announcing for all that can hear that we are on the verge of breakthroughs in our effort to understand our place in the scheme of things." If your defense of this doctrine is meant to offend, you have plenty of ammunition. If you want to question someone's religious beliefs, quote Wilson himself: "Although the manifestations of the religious experience are resplendent and multidimensional, and so complicated that the finest of psychoanalysts and philosophers get lost in the labyrinth, I believe that religious practices can be mapped onto the two dimensions of genetic advantage and evolutionary change."

Here is another quote from Wilson, guaranteed to provoke your favorite feminist: "Even with identical education and equal access to all professions, men are likely to continue to play a disproportionate role in political life, business, and science . . ." because this behavior seems to have a genetic origin. Whether the angle is politics, religion, or the battle of the sexes, sociobiology will be as much a battleground as a research area in the years to come.

RELATED TOPICS:

Altruism/Selfish Gene, Bioethics, DNA, Human Origins, Mutation, Origin of Life, Reproductive Technology.

FURTHER READING:

Sociobiology: The New Synthesis, Edward O. Wilson, Harvard University Press, 1975.

The Evolution of Human Sexuality, Donald Symons, Oxford, 1980.

STELLAR DEATH

DEFINITION:

Astrophysical theories which describe the characteristics of dying stars and explain how stars complete their life cycles.

WHAT IT REALLY MEANS:

One of the great unifying concepts of modern science is the realization that all forms of matter, both organic and inorganic, have definite life cycles. From the tiniest elementary particles that perish in trillionths of a second to huge stellar masses that exist for billions of years, all the corporeal stuff of the universe must heed the words of Ecclesiastes: "For every thing there is a season, a time to live and a time to die."

By human standards, stars may seem immortal, but in fact they are continually created and destroyed at a relatively rapid pace. New stars are born when giant old stars explode and send shock waves through clouds of interstellar gas, causing the clouds to gravitationally coalesce into protostars. Gravity continues to collapse embryo stars until thermonuclear fusion starts pushing outward, creating a stabilized hydrogen fireball as small as one hundredth the mass of our own sun or as large as one hundred solar masses. A few billion years after its birth, when the fusion process converts enough of the initial hydrogen mass into heavier elements, the star begins to die. How it dies is determined by its size.

With stars the size of our sun or smaller (which includes about 90 percent of all stars) death begins at the *red giant* stage: the supply of hydrogen available for fusion begins to dwindle. The star's hydrogen core is replaced by a helium core, which contracts and becomes hotter; helium fuses into carbon, continuing a progressively more complex synthesis of heavier elements, all the way up to iron. The added heat and radiation pressure from the core cause the outer layers of the star to expand, providing more surface area to dissipate internal heat. The star continues to expand and cool—thus becoming redder. If our sun were to become a red giant, it would expand out to the orbit of Mars, engulfing earth along the way.

When the thermonuclear pressure pushing outward drops below the level of the gravitational pressure squeezing inward, the star begins to collapse. The more it collapses, the more the force of gravity increases, until the matter in the star becomes hypercompact and stabilizes again as a *white dwarf.* A white dwarf of one solar mass would have a volume only one millionth that of the sun; one cubic centimeter of white dwarf would weigh about a ton. As dense as this matter is, it still consists of the same old elements found in the periodic table. The matter in a white dwarf is kept from complete collapse by the energy of its electrons. The white dwarf slowly cools until it no longer radiates visible light, and then it becomes a *black dwarf* forever.

A star possessing a mass between 1.4 and 2 solar masses continues to collapse past the white dwarf stage until neutron degeneracy takes over. The larger mass of these stars allows for a greater gravitational force and a correspondingly greater density—above ten million tons per cubic centimeter. Under this exquisitely intense pressure, matter itself is changed when the electron shells orbiting the atomic nuclei in the star begin to collapse. This supercollapsed star is known as a *neutron star.* The gravity at the surface of a neutron star is calculated to be about seventy billion times as strong as the puny force which keeps your feet close to the earth. The magnetic field of a neutron star outradiates entire galaxies in the X-ray and radio frequencies. A neutron star is one of the most bizarre speculations to emerge from astrophysics—its characteristics make it more like an outrageously heavy atom than a star.

As science-fictional as they may seem, neutron stars were predicted back in 1934 by Fritz Zwicky of Cal Tech. Thirty-three years later, after the development of radio astronomy, the first evidence that neutron stars might exist outside the imaginations of astrophysicists was received in the form of incredibly regular radio pulses from outer space. The signals were so precisely periodic that the first designation for their source was LGM—for "Little Green Men." Further analysis strongly suggested that the signals were being emitted by a *pulsar*—a rapidly rotating neutron star.

Pulsars are even weirder than the run-of-the-cosmos neutron star, because pulsars may rotate at speeds as high as one thousand revolutions per second; the rapid circling of the enormous magnetic field (a trillion times as strong as the terrestrial magnetism which commands compass needles) sends the pulsed radio beacons through space, in the manner of a rotating searchlight. "Pulsar quakes" have been theorized for the occasional jiggle in the ultraregular pulses.

A third scenario for stellar death occurs when a star is larger than two solar masses. When these massive stars die, there is no force which can stop their collapse. They simply continue to shrink to even smaller diameters than that of a baseball, strengthening their gravitational field as they collapse. By the time such a star is half the size of an average neutron star, its gravitational field is so strong that nothing, not even light, can escape. At this point, it becomes a *black hole.*

If a star is extraordinarily massive—at least five solar masses—it may go through a spectacular transition before finally dying out. A *supernova* occurs when a gravitationally collapsed star explodes instead of continuing to collapse. Supernovae are among the more notable celestial events because the exploding star's brightness may increase a million times in a period of days, and the energy radiated by a supernova in those few days might surpass our own sun's output over a million years. Although astrophysicists cannot explain the mechanisms by which supernovae are triggered, they do understand the crucial role they play in the formation of new stars. Supernovae explosions yield the heavy elements needed to produce viable planets such as the earth, and they also produce the tremendous shock waves that promote the formation of protostars. It is only through the death of some stars that others are created: "A time to break down, and a time to build up."

CONVERSATIONAL TACTICS:

If your conversational partners find it hard to relate life on earth to the distant deaths of massive stars, you could point out that our bodies are composed of ancient stardust. When the universe began with the big bang, the only elements available were the two lightest ones—hydrogen and helium. Then stars formed and nuclear fusion created more helium, and the pressures of gravitational collapse provided the energy to create carbon, nitrogen, oxygen, and the other building blocks of life. When the very rare supernovae occurred, they created the heaviest elements and blasted all these atomic essentials into space, where they drifted and clumped into solar systems and planets. Thus, every part of your body made of anything heavier than helium was cooked up in the heart of a star . . . and anything heavier than iron was once expelled from a stellar cataclysm!

RELATED TOPICS:

Big Bang, Black Holes, Cosmology, Entropy, Fusion, Galaxy, Quasar, Radio Astronomy, Sunspots.

FURTHER READING:

"Life on a Neutron Star," Robert L. Forward, *Omni,* 1980.
The Cosmic Connection, Carl Sagan, Dell, 1973.

SUNSPOTS

DEFINITION:

Darker, relatively cooler areas on the surface of the sun which appear in approximately eleven-year cycles and are believed to be associated with distortions of the sun's magnetic field.

WHAT IT REALLY MEANS:

In the early 1600s, when an Italian scientist named Galileo built a crude telescope and dared to report what he saw, he overturned established astronomical dogma, liberated science from the realm of metaphysical criteria, and opened the road to science based on observation and experiment. In the process, he also got himself in a famous battle with the Inquisition.

One of the most damaging pieces of evidence at Galileo's heresy trial was his 1609 observation that the sun had spots. The Church backed the ancient doctrine that the sun is a perfect sphere which orbits the earth. Thanks to Galileo and nearly four centuries of sun watchers since his time, we now know that the sun is far from perfectly spherical, that many of its blemishes are far larger than the earth and contain the energy of billions of H-bombs. Most recently, we have begun to learn how those dark spots on the sun are linked to devastating effects here on earth.

First, a few vital statistics about our sun. Sol is an average-sized, middle-aged star, and like all middle-aged stars is nothing more than a huge ball of gas—a sphere of hydrogen one million miles in diameter. With that much mass, gravity is so powerful that it causes the hydrogen at the sun's core to fuse, igniting the 30,000,000-degree thermonuclear furnace that is the sun's "engine." Every second, around half a billion tons of hydrogen atoms get squeezed until they become helium atoms, liberating an enormous amount of energy in the form of gamma rays; this radiation, in all its eventual transformations, is the primary source of energy in our solar system.

The center of the sun is so dense that it takes twenty million years for that energy to emerge from the sun's surface and radiate into space—which is good news for us, because that much gamma radiation would immediately kill all life. The reason it takes so long for this intensely penetrating radiation to work its way to the surface is that it is constantly colliding with particles, causing slightly weaker forms of radiation to ripple out and gradually losing penetrating power until it reaches us on earth mostly as the ultradilute form of radiation we know as visible light.

As the fusion-created energy journeys through the layers of the sun, it creates convection currents, which in turn affect the sun's powerful magnetic fields. These currents and fields, in conjunction with a peculiarity in the way the sun rotates on its axis, seem to be the causes of sunspots, flares, and other solar events. As the energy outflow approaches the surface, it drags the magnetic lines of force into a bulge, like pulling strands of taffy. These magnetic forces, thousands of times stronger than the earth's magnetism, are then wrapped around each other by the sun's uneven rotation. The sun rotates in belts, not all at once as on a solid planet. The faster rotation at the equator pulls and distorts the magnetic field into a very strong local field which eventually hemorrhages through the surface. The sunspots which occur in pairs appear to be the surface poles of those popped magnetic fields.

The current theory about sunspots postulates that they are cooler than the 6000-degree-centigrade surface because the locally intensified mag-

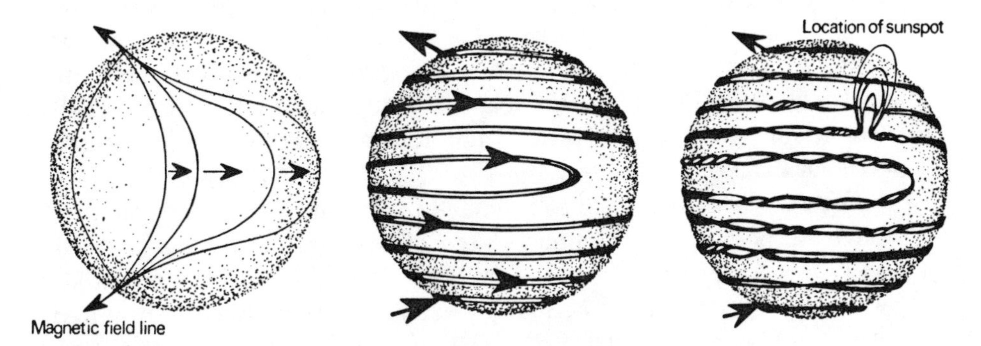

Sunspots—The Making of a Weather Maker

netic field blocks the flow of hot gas from the sun's interior. Sunspots appear in irregular numbers, ranging from none to hundreds at a time, in approximately eleven-year cycles. Although the connection between sunspot cycles and terrestrial weather is still conjectural, the appearance of large numbers of sunspots—known as a *solar maximum*—is always followed by intense solar activities which are now known to affect earth's atmosphere. The years 1979 and 1980 witnessed the strongest eruption of sunspots in the four centuries they have been observed, so solar scientists expect spectacular solar events to occur at unpredictable intervals during the 1981–84 period.

CONVERSATIONAL TACTICS:

It may seem strange that millions of tax dollars are spent to keep track of spots on the sun, since most people are hard-pressed to name any practical effect sunspots might have on their lives. It's a legitimate question: What can sunspots do for (and to) *you?* Take your choice of answers: a few of the known and suspected side effects of sunspots and associated flares, coronal holes, and solar winds are massive power failures, radio and radar blackouts that could throw everything from global navigation to the Strategic Air Command into communications chaos, radiation damage to passengers on polar flights, and occasional falling satellites. The most drastic of these forecasts is based on *solar flares,* which seem to be very closely related to sunspots.

A solar flare begins as an intense focus of radiation on the sun's surface near a sunspot; the fountain of X rays and gamma rays soon blows hot gas into space, like a superthin bubble on the sun's surface. When it pops, that plasma bubble propels billions of tons of hot molecules on a direct course toward the earth. The resultant shower of radiation has terrestrial effects which are only now being studied directly. One of the early casualties of the current solar maximum cycle was *Skylab,* which was literally pushed out of orbit by unforeseen atmospheric bulges caused by solar flares.

When one of those big solar bubbles pops, an X-ray storm hits the earth's atmosphere eight minutes later, blowing the electrons off oxygen and nitrogen atoms; this ionized gas layer above the earth absorbs radio-wave energy for a brief period, effectively "eating" radio-transmitted signals. The show is far from over after this potentially disastrous occurrence. An hour or so after the X-ray storm, a gale of fast protons from the sun's ruptured corona hits the outer layers of the atmosphere; most of these secondary particles are captured by the earth's magnetic field, but they tend to leak through at the poles, sometimes causing a radiation hazard to airplane passengers on transpolar flights. The protons also rattle the earth's magnetic field, causing electrical surges in power lines. During one outburst in 1859, this magnetic surge astounded telegraph operators, who found they could operate their machines without batteries! Imagine what a surge like that could do in today's "wired society."

In 1972, a solar flare had a consequence with ominous implications for our power-and-communication-dependent society. A transformer exploded in Canada and tripped safety breakers on the power grid all the way from Maine to Texas. The vulnerability of the power grid to local breakdown was dramatically illustrated by the East Coast blackouts of 1965 and 1977. Today, we are even more dependent on interlinked grids which have the unfortunate property of acting as huge antennae for these unpredictable magnetic storms.

RELATED TOPICS:

Coriolis Effect, Energy Alternatives, Fusion, Galaxy, Photosynthesis, Radio Astronomy, Stellar Death.

FURTHER READING:

Our Universe, Roy A. Gallant, National Geographic Society, 1980.

"The Sun Turns Savage," Dennis Overbye, *Discover*, November, 1980.

TECHNOLOGY ASSESSMENT

DEFINITION:

The systematic study of the effects on all sectors of society that may occur when a technology is introduced, extended, or modified, with special emphasis on any impact that is unintended, indirect, or delayed.

WHAT IT REALLY MEANS:

You don't need a Ph.D. in engineering to figure out that technologies don't always function exactly to specifications. All one need do is look at the headlines: recalls of products ranging from tampons to trucks; Hollywood strikes because of disputes about the potential money-making power of new video technology; a socket falling into a missile silo and a Titan's nuclear warhead being ejected; and arguments about the efficacy of everything from MX missiles to various medicines.

Episodes like these, and hundreds of others involving everything from acid rain to x-radiation from color TVs, have taught us that technology may do as much harm as good. They have also taught us that *caveat emptor*—"let the buyer beware"—is still good advice, though much less practicable in a highly technological civilization. Who has the resources to check everything from the abrasive qualities of a tampon to the effects of dropping weights on Intercontinental Ballistic Missiles? We all must depend on a surrogate, either the manufacturer or the government, to inform us about the consequences of technology. This is the role of technology assessment (TA)—to evaluate all the effects of a technology, both planned and unplanned, and to make an overall judgment about the merits of that technology.

Obviously, with disaster stories like those listed above, it is easy to see that TA is not foolproof. It is still more of an art than a science. However, it is a cheap shot to slam TA because of its failures. The fact is, we seldom get to hear about the successes because when a product is determined to be unsafe, it never reaches the consumer. (Remember—thalidomide was never marketed in this country because of TA.) Ultimately, this is the goal of the whole field—to decrease the number of nasty technological surprises. Unfortunately for us, the consumers, this is no easy task.

Simple product testing, a limited case of TA, is one thing. But how is one to assess fully the implications of a technology as complicated as television? Like all good systems engineers, technology assessors use a general checklist approach: What are the technical boundaries of the technology? What needs and markets is it designed to meet? What technological and social assumptions should be made about the future? What nonintended impact might the technology have? But knowing the questions is only half the battle. Who could have foreseen the effects of television on reading scores, or the balance of trade, or the family unit? Good technology assessors also have to be good futurists—a profession with a less than enviable track record.

Oscar Wilde said that experience is the name people give to their mistakes. Because the cost of experience is now so high, we should expect TA, imperfect as it is, to be a growth industry in our increasingly technologized future.

CONVERSATIONAL TACTICS:

Concerns about technology are hardly a new topic of conversation. As early as 1663, London workers tore down the new mechanical sawmills which threatened their jobs, and the Luddites turned anti-technology feelings into a hammer-wielding social movement in the early nineteenth century. What is new in the post-World War II era is the increased rate of technology acquisition and its increased potential for causing widespread harm. The little quiz below is designed to determine just how diligent a tech-watcher you've been in the last few decades. All these technologies found their way into the headlines as potential dangers. Match the numbers with the correct letters. A score of six or more qualifies you as a potential technology assessor.

1. AD-X2	A. Synthetic sex hormone
2. Cyclamates	B. Diet sweetener
3. Diethylstilbestrol (DES)	C. Disinfectant

4. Hexachlorophene
5. Krebiozen
6. Nitrilotriacetic Acid (NTA)
7. Project Able
8. Project Sanguine
9. Project West Ford
10. 2,4,5-T

D. Putative cancer cure
E. Detergent booster
F. Orbital communications system
G. Global communication transmitter
H. Orbiting space mirror
I. Battery additive
J. Herbicide

ANSWER KEY: 1-I, 2-B, 3-A, 4-C, 5-D, 6-E, 7-H, 8-G, 9-F, 10-J.

RELATED TOPICS:

Acid Rain, Appropriate Technology, Electronic Smog, Energy Alternatives, Recombinant DNA, Toxic Chemicals, Weather Modification.

FURTHER READING:

Technology and Social Shock, Edward W. Lawless, Rutgers University Press, 1977.

"Modern Technology: Problem or Opportunity?" *Daedalus,* Winter, 1980.

TOPOLOGY

DEFINITION:

The branch of mathematics that deals with transformations which are both biunique and continuous in both directions.

WHAT IT REALLY MEANS:

Most people tend to think of geometry as the mathematics of measurements—how long, how much area, how much volume, what size angle. Geometries such as Euclidean, which measure lengths and angles, are called metric. Topology is a nonmetric or nonquantitative geometry; it is concerned with the ways in which surfaces can be pulled, stretched, twisted, and transformed from one shape to another without breaking or tearing. A topologist is sometimes defined as a person who can't tell the difference between a doughnut and a coffee cup. To a topologist, they are the same one-holed object, which has simply undergone a transformation in shape.

Both the doughnut and the cup contribute to the language of topology, for another way of defining the topologist's task is "the process of finding out how many holes there are in an object." In the special topological sense, these mathematical holes are called *handles* and can be described in a precise way by appropriate sets of equations.

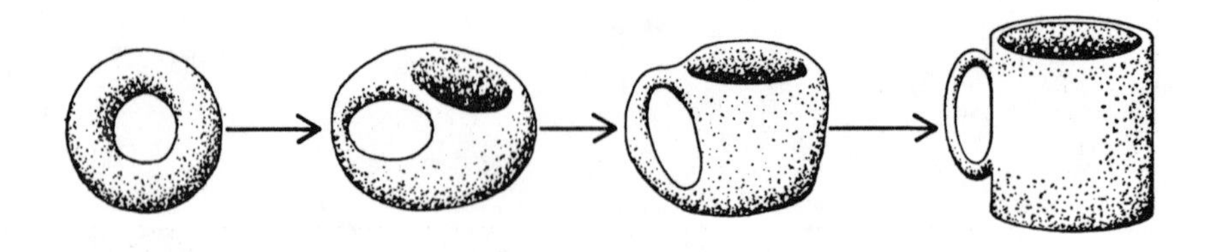

Topology—Doughnuts and Coffee Cups

Of what possible use is a branch of mathematics which would have us believe that a triangle is the same object as a square? Although most of topology's utility is found in the theoretical sciences, it has surprising consequences in everyday life. For example, drive belts wear out because of friction created when they move across the gears. If such a belt is twisted 180 degrees before the ends are joined, it will probably last longer, because both sides of the belt will take the wear evenly. In fact, as topology demonstrates, such a belt (called a *Möbius strip*) has only one side and one edge!

CONVERSATIONAL TACTICS:

The Möbius strip can easily be constructed with a scissors, a piece of paper, and a bit of tape. Cut a rectangular strip at least eleven inches long by about an inch wide. If you are at a Washington, D.C., party, or at any social gathering of professional bureaucrats, make it a *red* strip of paper and explain that you are going to demonstrate the paradox of "bureaucratic topology."

Bend that strip of paper into a loop and give the end a half twist before you join it into a circle. If you start on either "side" of the paper and trace a line along it with a pencil, you will arrive back at your

Topology—A G-Rated Strip

starting point without lifting your pencil, thus demonstrating that you have created a figure with only one side.

Announce that you have constructed a mathematical model of "red tape" and show how any way you try to cut it only makes it more complicated: If you cut the loop lengthwise into two halves, you end up with a single loop of red tape which is twice as long as the original; if you cut it lengthwise into thirds, you end up with two pieces of red tape inextricably linked together. The Möbius strip, like red tape, just gets longer the more you try to cut it.

Another practical demonstration of topology, and an amusing party wager, is the simple, if strenuous, method of taking off a vest without removing your jacket. This proves to your conversational companions that your vest was never really "inside" your jacket at all, in the topological sense.

RELATED TOPICS:

Analog Computer, Information, Mathematical Modeling, Relativity Theory.

FURTHER READING:

Mathematics, David Bergamini and the Editors of *Life*, Time-Life, 1963.

Men and Numbers, ed. James R. Newman, Simon and Schuster, 1956.

TOXIC CHEMICALS

DEFINITION:

Any poisonous substance, natural or synthetic, capable of producing harmful effects through physical contact, ingestion, inhalation, or radiation.

WHAT IT REALLY MEANS:

On the visceral level, we all understand the meaning of toxic chemicals. They represent the fouling of our own nests. Toxic substances are currently a hot issue in the scientific and political world, but they are also hot issues in your own neighborhood if there is an old chemical waste dump nearby, a leaky transformer on the pole outside your window, or a chemically defoliated patch in the forest behind your house. Simply put, many of the chemical substances which now surround us are poisons.

Two decades ago, all the talk on the tube was about "the miracles of modern chemistry" which were making our lives brighter, whiter, and so much more convenient. "Better living through chemistry" was a successful advertising slogan. That era ended with thalidomide, napalm, Agent Orange, asbestosis, and Love Canal. Where once it seemed to be the hope of the future, chemistry now seems to be viciously intruding on our lives. We live in a society which is totally dependent upon thousands of synthetic chemicals, yet every day another component of our technological environment is found to be carcinogenic, mutagenic, or brain-damaging. Although it has become a matter of survival, it is increasingly difficult to keep track of the poisons in our food, air, water, homes, and work environments.

Just as important as recognizing these dangers is the good sense to keep it all in perspective. There are currently more than seventy thousand synthetic compounds on the U.S. market, and less than one hundred have been indisputably linked with cancer in human beings. Even though several hundred more have been found to cause cancer in animals and an even larger number are believed to cause other types of harm, the vast majority of the components in our chemical environment seem safe.

Of course, the operative word in the last paragraph is "seem." Safe today may not be safe tomorrow, and better toxicological screening techniques, joined with expanded substance-testing programs, have a way of discovering new environmental threats. Ultimately, we should heed the words of an EPA administrator: "We look back on the Middle Ages and say, 'No wonder they had bubonic plague—they used to throw their garbage in the streets.' I just hope that in the year 2025 my grandchildren don't look back on this generation and say, 'No wonder they had problems—look at all the chemicals they carelessly introduced into the environment."

CONVERSATIONAL TACTICS:

Talking Tech's "Child's Garden of Toxic Chemicals" should do more than provide you with something to talk about next time someone mentions plutonium or PCB—this knowledge may help you make the decisions necessary to protect your life and health. Whether you are a stockholder in a chemical manufacturing corporation or the victim of a vinyl chloride spill, you need to understand the substances you are dealing with before you can intelligently decide what to do with them.

Asbestos: Unlike the newer, synthetic chemicals, asbestos is a naturally occurring mineral silicate which has been used in manufacturing for over two thousand years. It may also be the first environmentally hazardous industrial substance. As early as the first century A.D., Roman physicians noted a "sickness of the lungs," which we now recognize as asbestosis (a nonmalignant pneumoconiosis), in slaves who wove asbestos into cloth. Although the EPA set air emission standards for asbestos plants in the early 1970s, our society has not decreased its use of the mineral. About one million tons of asbestos are used each year in construction, textiles, and automotive systems. If you personally inspect your brake or clutch linings, beware the fine dust—it's asbestos.

DDT: Dichlorodiphenyltrichloroethane was the first great bug bomb. It emerged from World War II, when it proved effective in controlling malaria and typhus, to spawn the pesticide industry. For twenty years after the war, planes and spraying trucks literally soaked the nation in a variety of pesticides. During this time, there were a few reports about negative side effects, but nothing conclusive. In early 1962, the thalidomide tragedy gave the world a lesson in unintended side effects, and when Rachel Carson's *Silent Spring* hit the bookstores that fall, the nation was ready to listen.

Scores of scientific reports poured forth about the noxious effects of DDT, from the thin eggshells (which endangered bald eagles) to its putative ability to hinder photosynthesis in crops to the contamination of human mothers' milk. By 1969, the public clamor was so great that DDT was banned. What it left behind was a legacy of public education. First, there was an increased awareness of the food chain. We don't have to directly ingest poisons to get sick. Humans are at the top of every food chain, so if we eat animal or vegetable products which have poisons in their systems we'll end up with poisons in our systems. Secondly, insects which breed incredibly fast can build up an immunity to pesticides. Even if DDT were not banned, it would not be as useful today because it has accelerated the evolution of DDT-resistant superbugs. Third, and most important, we now realize that anything which poisons any part of the environment is likely to affect human beings sooner or later, one way or another.

DES: Diethylstilbestrol, one of the first synthetic sex hormones, was originally produced in the 1940s as a treatment for menopausal disorders. However, it became a commercial success in 1947 with the discovery that it could be used to fatten chickens—a small pellet of DES implanted under the skin of a young rooster acted as a chemical castrator, turning him into a plump capon. Both the beef and sheep ranchers soon followed the example of the chicken farmers and started using DES to fatten their stocks. By 1970, nearly 75 percent of all beef was produced with DES. Then, in 1971, a study linked rare vaginal cancers in young women with the use of DES by their pregnant mothers to prevent abortions and miscarriages. The threat of a "hormonal time bomb" was linked to the DES levels in animals. Although the link was never proven, the FDA and USDA banned the use of DES to fatten cattle at the same time (1973) that the FDA was approving the use of DES as a morning-after pill! As with other oral contraceptives, DES proved that, in matters of sex, it's difficult to improve on mother nature.

Dioxin: 2,3,7,8 tetrachlorodibenzoparadioxin (TCDD) is the deadliest human-made molecule ever assembled. As little as five parts per *trillion* in the diet causes tumors in rats, and five hundred ppt causes female monkeys to abort and die. Dioxin is found in a number of well-known herbicides: Agent Orange, 2,4,5-T, and silvex. Although all the evidence is not yet in, pregnant mothers near the forests of Oregon, which underwent treatment with these herbicides, now report increased numbers of miscarriages. The entire globe witnessed the potential horrors of ecological self-destruction in July 1976 in the Italian town of Saveso. An explosion of the nearby chemical factory spread a blanket of dioxin around the region. With a few days, pets and other animals were dying in the streets, and humans had symptoms such as skin rashes, diarrhea, and vomiting. Long-term effects and effects on the newborn remain to be seen.

PCBs: Polychlorinated biphenyls may sound like something from a chemistry laboratory, but you probably don't have to go much further than your own body to find the nearest concentration of this toxic chemical group. Studies have shown that over 90 percent of all

Americans have some PCBs in their bodies! Ten years ago, little was known about the medical effects of these oily chemical cousins to DDT, but a lot was known about their utility in an insidious range of industrial and consumer goods. Millions of gallons are still used in transformers as an insulating fluid—so if you see a high-tension pole with a metal box leaking fluid, steer clear and call your power utility. These versatile molecular inventions have also been used in fluorescent lights, typewriters, dishwashers, and paints.

In animal studies, the effects of PCBs are clear—deformity, disfigurement, sickness, and death in a range of mammals from mice to monkeys. As to human studies—well, long-term studies by definition take time, and the results are just now coming in. There is very little doubt in any investigator's mind that the outcome of human PCB-effects studies will be grim indeed. These molecules have the nasty resiliency that enables them to survive and enter the food chain. The FDA has had to confiscate hundreds of thousands of pounds of poultry, eggs, and fish because of PCB contamination.

Plutonium: Pluto, brother of Zeus, was the Greek god of hell itself, and the substance named after him has toxic attributes which can only be termed demonic. Plutonium was a crucial discovery in modern science, the agent of an irrevocable turning point in history, and it may well be the key substance of future apocalypse. You may not encounter it daily in your home or on your job, but there is not one person living on this planet whose future does not depend on how we use this devilish element. It has the ability to cause bone cancer in doses of one millionth of a gram, and when it is formed into the core of a nuclear weapon, it has the capacity to incinerate billions of human beings.

Unfortunately, the world's most dangerous element is rapidly becoming one of the world's most plentiful currencies. The U.S., Europe, the Soviet Union, and much of the developing world are becoming more and more dependent on a "plutonium economy" to generate nuclear power. If access to this substance could be absolutely controlled, there would be little danger, but it seems that alarming amounts of this supertoxin slip through the cracks in even the tightest security system. The nuclear-power and weapons community has a fondness for acronyms, and the handling/transportation of plutonium has spawned one of the most infamous—MUF, which stands for "*m*aterial *u*naccounted *f*or." There are already *tons* of plutonium "missing" from the system. In the 1990s, with tens of millions of pounds of plutonium in circulation, the increasing amount of MUF will pose a definite threat to all life on earth, either through deliberate nuclear terrorism or through massive fatal contaminations.

RELATED TOPICS:

Bioethics, Meltdown, Mutation, Technology Assessment.

FURTHER READING:

"The Pesticide Dilemma," *National Geographic,* February, 1980.
Laying Waste: The Poisoning of America by Toxic Chemicals, Michael Brown, Pantheon, 1980.
The Curve of Binding Energy, John McPhee, Farrar, Straus & Giroux, 1974.

UNCERTAINTY PRINCIPLE

DEFINITION:

A fundamental precept of quantum theory, asserting an inverse relationship between the measurable position and the momentum of an elementary particle.

WHAT IT REALLY MEANS:

The uncertainty principle, sometimes called the Heisenberg uncertainty principle, named after Werner Heisenberg, is a basic consequence of quantum theory. It means that it is impossible to measure precisely both the position and the momentum of the same moving particle simultaneously. This deceptively simple truth had two monumental implications for the physics of the twentieth century. First, it imposed limits beyond which science cannot accurately measure subatomic events. These limits are neither a function of the crudeness of our measuring devices nor of the remoteness of the events we seek to measure but are part of the way our minds comprehend reality. According to the principle, we cannot observe events on the subatomic level without changing them in some way. Second, and as a consequence of the first, we can no longer postulate an independent observer who is able to stand on the sidelines and watch nature run its course without disturbing it in the process.

Heisenberg's formulations also have serious implications for our "common sense" view of the world. In the previous century, people were shocked to discover that the solid objects of our everyday lives were actually composed of tiny particles whirling in huge spaces. Heisenberg reduced those infinitesimal but physically definite particles to probability fields and made the investigator a crucial part of the system under investigation. Overnight, the boundary between us and the rest of the world disintegrated. We were told, by the people who were the designated experts in these matters, that we can "know" about the universe only within prescribed limits of certainty and that our search for this knowledge has a role in setting those limits!

As a way of illustrating the logic of his principle, Heisenberg talked about a then hypothetical microscope powerful enough to resolve an electron moving in its orbit around a nucleus. Because the wavelength of visible light is too short to resolve electrons—like trying to roughly outline a cockroach by covering it with beach balls—a much higher-frequency, shorter-wavelength radiation must be used. That would be like covering the cockroach with grains of salt instead of beach balls—but the grains of salt would also have to be moving very fast. Heisenberg substituted gamma rays for photons and pointed out that in order for us to resolve the electron, the gamma ray would have to collide with it. However, this would irrevocably alter the electron's momentum, rendering it unmeasurable.

There is a blurring in the measurability of the subatomic world. If we try to increase the wavelength and lower the energy of the gamma-ray beam in order not to disturb the momentum, then we would have to give up our certainty of the electron's exact location. This uncertainty, in a fundamental sense, is built into the observer-electron system: when we focus on position, momentum is disturbed; if we focus on momentum instead, the position is blurred. The choices made by the observer determine the nature of the information obtained regarding a particle's position and momentum.

Uncertainty Principle—Two Views from the Bridge

It is tempting to regard our forced uncertainty as a technological artifact, as just another problem waiting for a more advanced theory or a more powerful machine, but this misses the whole point of the principle. It is not the power of our scientific instruments which is in question, it is the power of our minds. As Heisenberg himself stated: "What we observe is not nature itself, but nature exposed to our method of testing." We can no longer believe in a separate, external world beyond our minds, independently existing, waiting to be investigated. Our investigations, in a sense, determine the structure of our world. We cannot talk with any certainty about the "real things" behind the equations of physics—we can only talk about the correlation between the equations and our methods of measuring reality.

CONVERSATIONAL TACTICS:

One of the oldest and most provocative conversational ploys is the simple, naive question: "Why is the sky blue?" or "Why don't we fly *up*?" Not only do these verbal bombshells stop everyone else in their thought tracks, they give you a chance to properly explain a phenomenon most people think they understand but actually don't. Remember, though, that simple questions deserve simple answers; it might not be wise to start out simply, only to wander into the philosophical thickets surrounding Heisenberg's equation.

Unfortunately, modern science (and modern scientists) often seem incapable of asking simple questions; the whole enterprise of research has become so specialized that it is not unusual to find a scientist expanding on his academic microcosm ad nauseam without making sense to anyone else in the group. If you meet such a bore and sense that the audience is as bored as you, wait until the uninvited lecturer pauses to take a breath, then ask: "Why are atoms so big?"

This is guaranteed to give the pontificator pause to contemplate either the question or your sanity. While the professor starts to stammer a reply, volunteer that the uncertainty principle is the key to understanding the answer. Wait a few beats, then offer the following explanation.

Because atoms are composed of positive and negative particles, we would expect the opposing charges to attract each other, so that the protons and electrons would be right next to each other. In fact, the electrons orbit the protons at a relatively great distance, giving the entire atom a diameter of roughly 10^{-8} cm. On the atomic scale, this means that atoms are huge objects compared to their constituent parts. Why? According to the uncertainty principle, we can't know both the position and momentum of an electron; therefore, if the electron resided in the nucleus, we would know its position precisely, which in turn means that it would have to possess a very large momentum. Its necessarily high momentum would cause the electron to break away from the nucleus. As a kind of nuclear compromise, the electrons jiggle about in orbital clouds surrounding the nucleus. By the time you've made this clear, you will have steered "Professor Bore" away from the monologue.

RELATED TOPICS:

Blackbody Radiation, Consciousness, Entropy, Particle Accelerator, Periodic Table, Quantum Theory, Relativity Theory.

FURTHER READING:

The Tao of Physics, Fritjof Capra, Shambhala, 1975.

The Forces of Nature, P.C.W. Davies, Cambridge University Press, 1979.

VIRUSES

DEFINITION:

Submicroscopic particles which are capable of replication only within the cells of a susceptible host.

WHAT IT REALLY MEANS:

How small can a creature be and still be considered a living organism? If you were to construct your own simple life form, molecule by molecule, how many molecules would you have to add to the construction before you could call your creation "life"? At what point on the microscopic scale does chemistry become biology? With the age of biotechnology upon us, we're all going to be involved in the medical, biological, and philosophical answers to similar questions. A long scientific quest is rapidly approaching its goal: just as physicists have attempted to define the nature of matter by looking for "fundamental particles," biologists searched for a similar "fundamental particle of life."

Viruses seem to be just the organisms to help answer those questions, and we are sure to learn even more about viruses in the future, for they are as vital to medical science as they are intriguing to research biologists. Viruses may someday help forge the key to the ultimate secrets of life, but they are of more urgent interest as the causes of dreaded plagues. Influenza, rabies, polio, smallpox, and other viral diseases have killed hundreds of millions of people, and the medical breakthroughs of vaccination and immunization have saved millions of people who are alive today only because of aggressive research into viral infections. And, as if the subversive little life forms weren't virulent enough, the way viruses work appears tantalizingly similar to the cellular transformations involved in cancer.

Clearly, there is far more at stake in the virus game than simple scientific curiosity, but without pure research, we would never have progressed as far as we have. If you have ever peered through the eyepiece of a microscope, you'll recognize the sense of wonder, the mental leap into the dimensions of fantasy, that motivated the early biologists. For four hundred years, biology students have squinted in amazement at the same awe-inspiring sight—the frenetic, fantastic world of all those unsuspected creatures which populate a single drop of pond water. Ever since the first Dutch lens grinders late in the sixteenth century made optical instruments capable of resolving these "animalcules," naturalists and medical doctors have wondered how far into the land of the infinitesimal these worlds of life within worlds of life could extend themselves.

Although viruses are small, even on the submicroscopic scale, their devastating effects upon much larger organisms, like people, have been known for a long time. A century before the electron microscope made them visible, Louis Pasteur experimented with viruses by trying to capture them in porcelain traps. Even without understanding anything about their structure or function, Pasteur knew that something capable of passing through the finest filter was still capable of producing infection or conferring immunity: when he passed the saliva of rabid animals through porcelain filters capable of trapping the tiniest known microorganisms, the filtered saliva was still lethal when injected into other animals.

Jump forward a hundred years to the 1950s, when the young science of molecular biology was able to apply more powerful tools in order to crack the secret of these potent, nearly infinitesimal, maddeningly unclassifiable creatures. A special kind of virus known as a *bacteriophage* —bacteria eater— had become a very important animalcule around the laboratory. Like human, other animal, and plant cells, even the microscopic bacteria are susceptible to viral infection. By studying the effects of viruses upon certain bacteria, biologists were able to zero in on what they are and how they work.

We now know that a virus is nothing more than a single strand of either RNA or DNA, encapsulated in a special protein coating; it consists of a head, which contains the nucleic acid core, and a tail, which plays an important role in viral reproduction. The fact that they contain a nucleic acid core is what makes viruses lifelike—DNA is, in fact, the "fundamental particle of life," the point at which a complex molecule becomes alive. The aspect of viruses which makes them *unlike* most other life forms is their inability to replicate outside a host cell belonging to a different organism.

The mechanics of bacteriophage infection are now well understood:

special chemicals on the virus protein coat are attracted to, and have the ability to break down, certain molecules on the surface of target cells. Once the tail of the virus has made contact and penetrated a host cell, it literally injects its nucleic core into the cell, leaving the coating outside the cell membrane. Once inside the invaded cell, the viral DNA acts very much like an army of occupation: it sends out messages to the host cell to cease production of the proteins necessary for that cell's life processes, then subverts the cell's own chemical machinery by instructing it to produce new virus heads, tails, and cores; the virus parts assemble themselves within the cell, then cause an enzyme to be secreted which bursts the cell walls, causing hundreds of new viruses to spew forth.

What is true of viruses is very close to what is true of *all* life—in a sense, we are all just collections of host cells for our DNA messages. A virus is nothing more than *information* with a thin protein coating. Life itself is a message, an information-dense pattern of chemical reactions encoded in the molecular arrangement of our nucleic acids. In terms of evolution, we are born to pass on that message, which weaves its subtle variations long after our deaths and the deaths of our grandchildren. Like a grain of wheat, a dolphin, or a human, a virus is a message from one part of the gene pool to another.

CONVERSATIONAL TACTICS:

If you want to steer your conversation into outer space, consider the interstellar plague theory. In *Diseases from Space*, the eminent cosmologist Sir Fred Hoyle and the equally eminent biologist Chandra Wickramasinghe argue that life began on earth four billion years ago because comets carried prebiotic substances to our planet and micrometeors later rained life itself onto the prepared surface. They speculate that interstellar clouds of influenza viruses may still rain from space at unpredictable intervals, perhaps causing those mysterious and deadly worldwide outbreaks of new infectious strains.

Even more controversial and speculative is the suggestion of two Japanese scientists. The next time your conversational companion mentions that he "caught a virus," suggest that it might be a message from extraterrestrial intelligence. Hiromitsu Yokoo and Tairo Oshima suggested in 1970 that a recently decoded virus known as øx-174 could be a clever message from space! When British researchers decoded the

DNA message of this bacteriophage, they were surprised to learn that the message could be read three different ways, for no known biological reason—like an ambiguous phrase that can mean different things, depending on how it is punctuated. The Japanese investigators suggested that such a viral code would be an ideal way to transmit messages through space. Conceivably, viral messengers could survive the long journey because some viruses have mechanisms to repair damage caused by ultraviolet radiation at wavelengths not experienced on earth. The messenger could replicate when it reached a suitable environment, transmitting extragalactic greetings not through radio-telescope programs but through quasi-living organisms.

Virus—Up Your Nose with Extraterrestrials

RELATED TOPICS:
Altruism/Selfish Gene, Bioethics, DNA, Exobiology, Mutation, Origin of Life, Recombinant DNA.

FURTHER READING:

Diseases from Space, Fred Hoyle and Chandra Wickramasinghe, Harper & Row, 1980.

The Eighth Day of Creation, Horace Freeland Judson, Simon and Schuster, 1979.

WEATHER MODIFICATION

DEFINITION:

Any human-caused influence, whether intended or not, which has the effect of altering the weather or longer-term climate.

WHAT IT REALLY MEANS:

For our entire history, the human race has been vulnerable to variations in weather and climate. Now there is growing evidence that the weather and climate may also be intensely vulnerable to human technology. Ways in which humans affect climate and weather include the massive amounts of heat that urban centers discharge into the atmosphere; gaseous wastes such as sulfur oxides which pollute the air and return in the form of acid rains; the release of carbon dioxide and other infrared-absorbing materials that occur when fossil fuels are burned and which help to warm the climate through the "greenhouse effect" (i.e., CO_2 allows sunlight in but prevents some heat from radiating back into space, thereby creating a warming filter in the same way the panes of a greenhouse trap heat); the threat to the ozone layer caused by the injection of oxides of chlorine and nitrogen; the increasing amount of aerosols (e.g., dust and particulate matter) that also accompany the combustion of fossil fuels and that alter the radiative properties of the atmosphere, thereby affecting the heat balance. If that staggering list of unintended effects isn't enough, there are now *intended* effects, various attempts to deliberately modify the weather through cloud seeding, fog suppression, and other techniques.

Although modern civilizations have devised ways to shield themselves from "the cold, cruel rain," those shields—homes, central heating—require huge amounts of energy, which, we now learn, produce more "cold, cruel rain" and other, more far-ranging climatic problems. Weather and climate are affected in two distinct ways by energy production. The first is *thermal pollution.* Since all processes that use energy release heat, the more energy consumed, the greater the amount of thermal pollution. It is now established that large urban centers form what are known as "urban heat islands" which are warmer in both summer and winter than the surrounding suburbs. These islands propel air upward at a greater than normal rate; at the same time, some of the small bits of dust emitted by industry or automobiles are carried aloft, where they serve as nuclei for cloud droplets. These mechanisms suggest that urban centers may influence precipitation patterns, both over the city and for some distance downwind.

The second way in which climate is affected by energy production is global in scope, and potentially far more dangerous. It is well known that increases in atmospheric concentration of carbon dioxide and aerosols which accompany the combustion of coal, oil, and gas alter the radiative properties of the atmosphere. The amount of CO_2 in the atmosphere seems to have increased about 10 percent in the past fifty years, and projections have it doubling current levels by 2050 and quadrupling them by 2100.

The "CO_2 crisis" is extremely significant because some mathematical models of the climate indicate that global surface temperatures will increase as CO_2 levels increase. For each doubling of CO_2, the models project a 2 to 3 degree (centigrade) average global temperature rise. This could mean a temperature increase of four to six degrees by the end of the next century. While nobody knows enough yet to accurately predict the consequences of such a rise, similar shifts in the past are linked to large-scale climate changes. What happens to coast cities if polar ice caps melt?

The situation regarding aerosols is less clear. Aerosols arise directly from injection of dusts by coal-burning operations and other industrial sources, as well as from slash-and-burn agricultural practices. There is also evidence that they are created photochemically in the atmosphere from unburned hydrocarbon fuel vapors and sulfur dioxide under the influence of solar ultraviolet radiation. Measurements at a number of sites around the world have shown a steady rise in the aerosol content of the lower atmosphere over the past few decades.

Although the concentration of aerosols is increasing, their effects on the climate are uncertain. Over land with a moderately high surface reflectivity, typical lower-atmosphere aerosols tend to warm the atmospheric column in which they lie and at the same time decrease the amount of solar radiation reaching the surface. Over the oceans, which have low reflectivity, aerosols tend to cool, since relatively more sunlight is reflected back to space when aerosols are introduced over a dark surface. Aerosols not only affect the radiation balance, but they also influence the formation of clouds and precipitation. Scientists are just now beginning to understand the total effect that aerosols may have on the climate, but it is clear that increased use of fossil fuels will certainly alter the weather, whether or not the consequences are fully understood.

If the "greenhouse effect" is not yet a cause célèbre of environmental groups, the same can't be said of protection of the ozone layer. During the 1970s, the ozone layer and human effects on it was twice a topic for heated public debate. First was the debate over the SST (supersonic transport) and its potential effect on the stratosphere. Then there was the issue of aerosol cans and the impact of the propellants they used.

Why was there such a clamor over an air stratum? The layer of air closest to the earth, ranging outward about eight miles, is the *troposphere.* This is followed by the *stratosphere,* and it is in this layer that the ozone resides. Unlike the troposphere, the stratosphere has very little vertical mixing of air and no precipitation. This means that the region is very sensitive to the cumulative effects of small amounts of contaminants.

Within the stratosphere are small concentrations of the gas ozone. Ozone is important to life on earth because it absorbs the wavelengths of solar ultraviolet radiation that can destroy cells and cause skin cancer. Any change in the amount of ozone in the stratosphere will affect its temperature, which will alter the circulation of the lower atmosphere and ultimately modify the climate down here at the bottom of all those layers.

The balance between ozone formation and destruction can be shifted by the addition of chemicals such as the oxides of chlorine and nitrogen. The oxides of chlorine come mainly from aerosol propellants, and the oxides of nitrogen originate mainly from nitrous oxide flowing from metabolism in plants and from effluents of stratospheric aircraft (e.g., SST). Regarding the SST, new information indicates that nitrous oxide additions to the lower stratosphere might increase the amount of ozone. This is a reversal of earlier projections. However, the original effects of chlorofluoromethanes (found in aerosol propellants) have been found to be twice as destructive of ozone as previously thought. It is clear that more research is needed. While one batch of engineers dreams up new and needed technologies, we had better have another group of scientists trying to determine the effects our inventions may have on the air we breathe.

Not all of technology's effects on the climate are unintended. In 1976, fifty-nine U.S. projects, most of them for cloud seeding, were listed in the World Meteorological Organization's "Register of National Weather Modification Projects," yet this number is a one-third decrease from 1974, and weather modification has yet to meet with the success originally predicted for it. While some of the reasons for this are technical, "weather mod" is a classic example of a technology far in advance of its attendant social, political, and economic effects.

If one group causes it to rain on their property and another group suffers a drought some distance downwind, can the second group sue for the diversion of precipitation? If one group seeds the clouds to prevent hail, can they be sued for reducing the rainfall? If a state engages in weather modification for the benefit of its citizens, can an individual enjoin the state on grounds that he or she is being injured? If two states disagree over cloud seeding, should the federal government intercede? These questions arise because there is not yet an answer to the question: Who owns the weather?

CONVERSATIONAL TACTICS:

Weather is the all-time favorite small-talk topic. It is also possible to elevate weather to the status of "big talk," the kind of topic to which people pay more than perfunctory attention. Next time the conversation lags and someone lamely mentions that "the weather was nice today," counter with the blood-chilling tale of The Coming Ice Age.

Ice ages are not necessarily events which happened tens of millions of years ago. The earth's climate oscillates between lengthy periods of great glaciations and other, shorter, warmer periods when the glaciers recede to the polar ice caps. In fact, the last great ice age ended only seventeen thousand years ago, and a little ice age occurred from the late fifteenth through the early nineteenth century. All the scientific evidence, everything from growing ice packs to lower temperature readings and southward-migrating armadillos, now indicates that since 1940 the climate has been getting colder. What is worse, the present cooling trend will be felt mostly in the temperate zones (where most of the world's population lives), and not in the tropics. The national sport of

Guatemala won't become skiing, but the heating bills from Boise to Bucharest are going to soar.

To be sure, no one knows if New York City will be covered with ice in a few hundred years or whether we're only on the precipice of another little ice age. It is also unclear just how modern technology might be affecting these long-term climatic cycles. Industrial exhausts are adding dust to the atmosphere, which has a cooling effect, but burning fossil fuels creates a greenhouse effect, which warms the atmosphere. There can be no doubt, however, that we're in the grip of a long-term cooling trend. That famous swamp land in Florida may soon start to look like a good investment.

RELATED TOPICS:

Acid Rain, Bioethics, Coriolis Effect, Photosynthesis, Sunspots, Technology Assessment.

FURTHER READING:

"What's Happening to Our Climate?" *National Geographic*, November, 1976.

Science and Technology: A Five-Year Outlook, National Academy of Sciences, 1979.

ZENO'S PARADOX

DEFINITION:

A series of arguments advanced by the pre-Socratic philosopher Zeno of Elea, designed to prove that motion is impossible.

WHAT IT REALLY MEANS:

Suppose that a friend challenges you to run the length of a football field. Just as you're about to begin, the friend advances the following argument: "It is impossible to traverse the field. Before you reach the other goal line, you must reach the fifty-yard line, and before that the twenty-five, and before that half the distance, and so on. Since space is infinitely divisible, and any finite distance must contain an infinite number of points, you will never reach the opposite goal line because it is impossible to reach the end of an infinite series in a finite time."

One reaction to this argument might be termed the Dr. Johnson response (Johnson is said to have tried to disprove Bishop Berkeley's theory that the only reality is our sense perceptions by kicking a rock and proclaiming: "I refute it thusly."). In other words, just run to the other goal. Such a response would not have fazed Zeno. He would have simply responded that it only appeared as if you had moved, but that, in

Zeno's Paradox—Zeno's Embarrassment

reality, you did not. He would have chuckled at your attempt to present physical evidence against a logical argument. He would have insisted that if you believed motion to be possible, you must demonstrate the fallacy in his argument.

In Zeno's time, his argument was a true paradox (the word comes from the Greek *paradoxos*, meaning "unbelievable," literally "beyond that which is thought"). Using just a few principles which all his contemporaries agreed to be true, Zeno was able to prove an unbelievable result. Today, modern science has demonstrated a way out of the paradox. Can you spot it? Zeno assumed that an infinite succession of intervals must add up to an infinite interval. We now know, thanks to the work of eighteenth- and nineteenth-century mathematicians, that certain infinite series known as convergent series add up to a finite amount. However, we should not be too smug. Zeno's argument stood for two thousand years.

The history of paradoxes in their literal rather than modern, exclusively logical, sense is a history of scientific progress. The solution to the "scandal of the ultraviolet," the explanation of the absolute speed of light, and the proof of a hierarchy of infinities were results "beyond that which is thought." Science and technology offer not just a way to manipulate the world but a way to perceive and conceptualize it. If this book has helped you to better understand the cosmos, to solve some of its paradoxes, then we have succeeded in the most important part of teaching you to Talk Tech.

CONVERSATIONAL TACTICS:

Not all paradoxes have been solved. Here are two modern paradoxes, only one of which has a reasonable solution. Can you guess which one is solvable and its solution?

1. The village of Hirsute has only one barber; this barber shaves all and only those men in the village who do not shave themselves. Does the barber shave himself?
2. A word is considered autological if it is self-descriptive (e.g., "English" is English, "tiny" is tiny) and heterological if it is not self-descriptive (e.g., "French" is not French, "long" is not long). Is the adjective "heterological" autological or heterological?

(Answer: The barber paradox is solvable. There can be no barber, because in giving the description you have stipulated an impossible world. There is no answer to the second paradox . . . yet.)

RELATED TOPICS:

Gödel's Incompleteness Theorem, Mathematical Modeling, Topology.

FURTHER READING:

The Presocratic Philosophers, G. S. Kirk and J. E. Raven, Cambridge University Press, 1964.

The Ways of Paradox and Other Essays, W. V. Quine, Random House, 1966.

EPILOGUE: A Puzzling Synthesis

Altogether they puzzle me quite, they all seem wrong and they all seem right.

ROBERT BUCHANAN

Science is a great game. It is inspiring and refreshing. The playing field is the universe itself.

—ISADOR I. RABI

ACROSS

1. It proved $E = mc^2$
23. Man's name (var.)
24. Gauguin's hideout
25. Glycerins
26. Peruvian city
27. Basic biological unit
28. Capital of Manche
29. Canine command
31. Type of pain theory
33. Alternative energy source?
34. Chase and Wharton have them
35. Affirmatives
38. Planck's discrete units
44. Musical note
46. Repression of CO_2 reflex
52. Majority
54. Office of Science and Technology Assessment
55. RNA nucleotide
56. There the same (Latin abbreviation)
57. Alternative energy source?
58. Double curve
59. Nanocuries
60. Type of armament
61. Pertaining to (Latin)
62. Keeper of ultimate measurements
63. 135 down does this to light
64. Interjection, for Einstein
65. Solar as well as logical
68. Electrical measure
69. Tumor-causing disease
71. Internal (abbreviation)
72. Wing part
73. Schumacher's subject
74. Weather causers
75. Britain's Mt. Palomar (abbreviation)
76. Fossil suffix
78. Sun prefix
79. Radioactive element
80. Scientific subject (abbreviation)
81. As shown below (Latin abbreviation)
82. Young sheep
85. Greek's "male" element
87. Powerful nuclear force
91. 3.14159
93. Auricular
96. Spanish river

97. Richard Speck's problem?
98. Nebraska city
99. Blood factor
101. Shed
102. Man's name
103. Sign of dreaming
104. Reversing electricity
105. Scientific evaluation
106. One, for Heisenberg
107. Pitcher
108. Charged atoms
109. Med——— boards
110. Roman coin
112. 100 ergs
113. King——— cigarettes
114. Young boy
115. Author of 184 across
119. Benign tumor
120. Type of daughter
122. Neutron absorber
123. Begins anew
124. —— Meson
126. Owns
128. Sun god
129. Small thicket
131. Raid
132. Fireman's friend
134. Call out
136. Negative-positive (Spanish)
138. Facial expression
140. French city
141. Color perceptors
142. Implementer
144. Tin
145. All there is
146. Taxonomic group (abbreviation)
147. Geometric solid
148. Ingredient in glowing paint
149. Three
150. Dr. Thom's idea
167. Man's name
168. Alchemists' goal
169. Marijuana's psychoactive agent
170. The nondigit
171. Star in Lyra
172. Diagnostic procedure
173. Reproductive technology
174. National Science Foundation
176. Air prefix
178. See 93 across
180. Past bioethics violator
181. Rave's partner
183. Nobel's invention
184. See 1 across

DOWN

1. Speedy particle
2. Semiconductor symbol
3. Kepler's orbits
4. Fossil fuel
5. Inert gas
6. Professor Hahn
7. Terrorist group
8. Get away!
9. C/D
10. —— tu Brute?
11. Tomb
12. Type of matter
13. Lightest solid element
14. Football lineman
15. Word in proof
16. Power delivered
17. Association for Scientific and Technological Evaluation
18. Integrated circuit
19. Doo's partner
20. Type of disease
21. ——— tse fly
22. Boated
28. State
30. Not ultraviolet
32. Not the subway
36. Quick exit
37. Window parts
38. Universe's oldest objects?
39. Sumerian city
40. Scientific society
41. Montana technology group
42. Multiplication
43. Most abundant metal
44. Fiery animal?
45. Fraud
46. Type of literature
47. Albacores
48. Book identifier
49. Exchequer

50. Interjection
51. Law of thermodynamics
52. Traveling residences
53. Mailed out
66. Radioactive isotope
67. Greek letter
69. Nutmeg state
70. Similar to dysprosium
72. Alternative energy source?
77. Atomic particle
83. Overseas
84. Nobel winner in 1954
85. British philosopher of science
86. Mutual association
87. Yoga masters
88. Tac's partner
89. Neural ———
90. Igneous rock
91. Olfactory attractant
92. Mother of Romulus
94. Ticketed cars
95. Women's names
97. Roentgen's discovery
100. Greek male prefix
110. Starlike
111. Solid, liquid, gas
115. Comparative suffix
116. City in Equatorial Guinea
117. Ireland
118. Nanosecond
121. Type of boat power
122. Our genus
125. Employer
126. Carry
127. Short interval
129. English union
130. Dutch airline
132. Water
133. Direction
135. 184 across is about this
137. Strong forward flow
139. Primeval prefix
141. Measure of light
143. Type of knight
150. Burn
151. ————-deucy
153. Black Sea bay
154. Photoelectric element
155. Photosynthetic machine
156. Horse's color
158. Deadly element
160. Malevolent
161. Lachrymal fluid
162. Liquid metal
163. Title of nobility
165. Lease
166. Abominable snowman
175. Rusty element
177. Regimental order
179. Rhenium
180. Most populous state
182. College degree
183. Palmlike plant

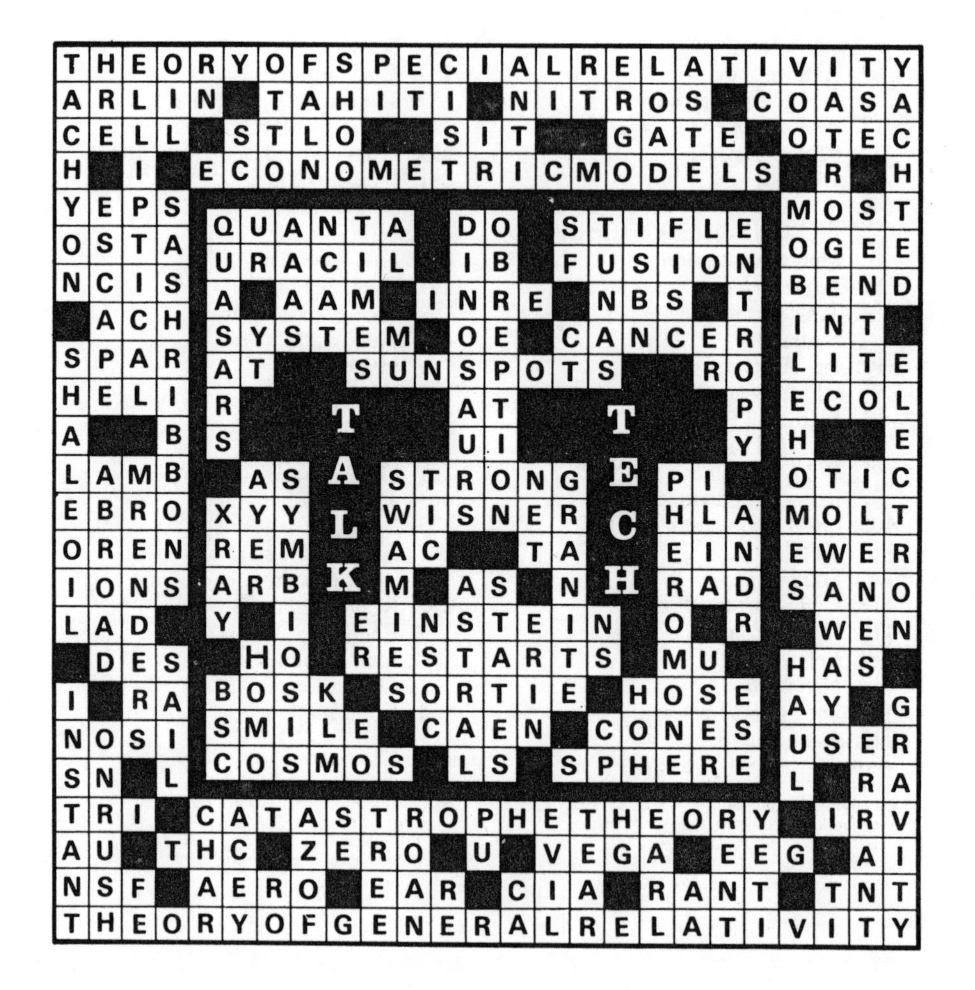

TALK
TECH

INDEX